The Nature of Time

The Nature of Time

Ulrich Meyer

CLARENDON PRESS · OXFORD

OXFORD
UNIVERSITY PRESS

Great Clarendon Street, Oxford, OX2 6DP,
United Kingdom

Oxford University Press is a department of the University of Oxford.
It furthers the University's objective of excellence in research, scholarship,
and education by publishing worldwide. Oxford is a registered trade mark of
Oxford University Press in the UK and in certain other countries

First Edition published in 2013

British Library Cataloguing in Publication Data
Data available

ISBN 978–0–19–959933–2

To Ombretta

PREFACE

If you want an exciting subject made dull, there are few better ways than having academics write about it.

The Economist, 6 May 2006, p. 84

The philosophy of time has a long history, but not a glorious one. Ancient and medieval authors often concentrate their efforts on complaining about the puzzling nature of their subject matter, without doing much to address the underlying issues. In my view, the reason for this slow progress is not that there is something paradoxical about time itself, as is sometimes suggested, but that our predecessors lacked the appropriate tools with which to attack the question. Like worries about the nature of the infinite, which were only resolved in the nineteenth century, I believe that a satisfactory account of the nature of time requires resources of modern logic that were unavailable to our forbears.

This book promotes a *modal theory of time*. Its central thesis is that instants of time are more similar to the possible worlds in modal logic than they are to places. While rival *spatial* accounts of time can be formulated within the framework of a standard extensional logic, modal theories usually take the form of a tense logic that spells out temporal distinctions in terms of conceptually primitive tense operators. Even though tense logic has some medieval precursors, it only emerged as a rigorous formal discipline with the pioneering work of A. N. Prior in the middle of the twentieth century.[1] My account is inspired by Prior's work, but I think that he did modal theories of time a disfavour by conflating them with two other theses, both of which I want to reject.

Like many of his contemporaries, Prior thought that tense logic could do double duty as a metaphysical theory of the nature of time and as a linguistic theory of verb tense. One of the reasons why modal theories subsequently fell out of favour is that it became clear that tense logic fails as a linguistic theory. While one could imagine a language that expresses temporal distinctions in terms of tense operators, this is not how natural languages actually deal with this issue.[2] I am happy to

[1] Prior (1957b; 1967; 1968c); see also Flo (1970) and Ørstrøm and Hasle (1993).

[2] See Meyer (2011: sec. 4) for a brief survey of the main counterexamples.

grant this point, but I also think that tense logic does much better as a metaphysical theory of time than is usually allowed. That is the part of Prior's project that I want to revive.

The second issue on which I diverge from Prior concerns the status of the present. Prior believed that any theory of time that is cast in terms of tense operators is *eo ipso* committed to an objective difference between the present moment and past and future times. I think this is doubly mistaken. Not only is there nothing metaphysically special about the present, modal views of time need not claim otherwise. On the account of instants I defend, the present is a time like any other.

There are as many modal theories of time as there are systems of tense logic. I favour a very weak system that imposes no structural constraints on the time series. For example, it is agnostic about whether the time series is dense or discrete, whether it has branches, or whether it loops back onto itself. On such a *minimal theory*, most of what other accounts regard as part of the nature of time itself are taken to reflect contingent features of what is happening within time. This is important for reconciling our metaphysical views with the best physics of time. Because my modal theory does not single out the present as metaphysically special, and because it imposes no constraints on the structure of the time series, either, it is easily reconciled with the theory of relativity.

My defence of this theory of time is organized in a fairly straightforward manner. Chapters 2 and 3 explain what is wrong with spatial theories of time; Chapters 4–8 develop my own modal view of time; Chapter 9 is about the present; Chapter 10 about the flow of time; and Chapters 11 and 12 discuss the theory of relativity. To help readers find their way around, there is a table of contents and a subject index. Instead of a person index, there are page numbers at the end of each bibliography entry that indicate where the item in question is referred to in the main text. To find out what I have to say about Prof. X, readers should therefore consult the bibliography, not the index.

Parts of the material in this book were presented at a number of different venues, including two conferences on the philosophy of time at the Università degli Studi di Bergamo in 2007 and at Wake Forest University in 2010. I would like to thank all the participants for very helpful comments and discussions. I also benefited from numerous formal and informal interactions with the worldwide community of philosophers of time. Some of them might not have realized the extent of their impact, but it is gratefully acknowledged: Yuri Balashov, Adrian Bardon,

Craig Bourne, Craig Callender, Thomas Crisp, Barry Dainton, Natalja Deng, Yuval Dolev, Heather Dyke, Katherine Hawley, E. J. Lowe, Ned Markosian, Trenton Merricks, Kristie Miller, Josh Parsons, L. A. Paul, Simon Prosser, Thomas Sattig, Steven Savitt, Jonathan Tallant, Michael Tooley, Stephan Torre, Nathan Oaklander, Steven Savitt, and Dean Zimmerman. Steven Savitt's graduate students at the University of British Columbia read a penultimate draft of the manuscript and are responsible for many improvements in the final version. Thanks are also due to two anonymous reviewers and to Peter Momtchiloff and his colleagues at the Clarendon Press, who were remarkably patient when I took much longer completing the manuscript than I had promised. Finally, a special debt of gratitude is owed to Judith Jarvis Thomson for crucial support in the early stages of the project—and for everything else.

Ulrich Meyer

Bologna
May 2012

CONTENTS

1

INTRODUCTION

THIS book is a long discussion of a single question: What sort of things are instants of time? This is not the only question that philosophers of time worry about, and for much of history it was not even a prominent one. Mediaeval philosophy was dominated by the question of how God relates to time and, more recently, the focus has shifted to the differences between past, present, and future, and to various attempts at reconciling accounts of the nature of time with the theory of relativity. These are important issues, but I shall argue that the nature of instants is *the* central question in the philosophy of time. Once it has been answered, all of these other problems easily fall into place.

Many philosophers think that times are similar to locations in space. Such *spatial accounts of time* can be distinguished further, depending on which of the two main views about the nature of places they take as their model. Spatial substantivalists, such as Isaac Newton, claim that space consists of metaphysically basic points that form a three-dimensional spatial manifold $\mathcal{M}$.[3] In order for them to exist, material objects must bear the spatial location relation L to some point on the manifold, but the converse is not true. Though it does in fact contain a great number of them, it is in principle possible for the manifold $\mathcal{M}$ to be devoid of material objects. *Temporal substantivalists* think that time has a comparable structure, and that it forms a one-dimensional temporal manifold $\mathcal{T}$ of metaphysically basic time points. To say that an object a exists at time t would be to say that it bears the exists-at relation E to the corresponding point on $\mathcal{T}$. As in the spatial case, it is impossible for material objects to exist without existing at some time or other, but in principle possible for time to be completely empty.

[3] I am using 'manifold' as a neutral term to refer to the collections of points postulated by spatial and temporal substantivalists. Since one of these manifolds is meant to be a theory of physical space and the other a theory of time, calling both of them 'spaces' would be confusing. I am thus using 'manifold' more liberally than what is custom in differential geometry, where it is reserved for spaces that are locally isomorphic to $\mathbb{R}^n$. As I understand spatial substantivalism, it is compatible with space being discrete, in which

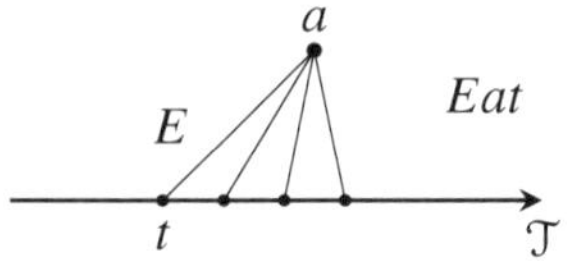

Figure 1.1. Temporal substantivalism

Relationists about space, such as Gottfried Wilhelm Leibniz, think that spatial substantivalists have the dependence between space and material objects backwards. Rather than material objects requiring the existence of space, they believe that space requires the existence of material objects. On their view, there is no spatial manifold $\mathcal{M}$ that exists independently of its contents. All there is are material objects that stand in spatial relations to one another. Places are just the positions in the order that spatial relations impose on material objects. Since there is no space to be abstracted unless there are at least some material objects, relationists conclude that space cannot be entirely empty. *Relationists about time* argue, in a similar manner, that temporal substantivalists are wrong about the dependence between objects and times. They deny the existence of the temporal manifold $\mathcal{T}$, and argue that time requires the existence of material objects. According to one popular way of spelling out this proposal, all there is to time are events that stand in the relations of temporal overlap and precedence to one another. Times mark the positions in this order and can be thought of as classes of pairwise overlapping events. In the case depicted in Fig. 1.2, time t is the set of events $\{e_2, e_4, e_5\}$, and t' the set $\{e_6\}$.

Instead of adopting either of these spatial views, I want to defend a *modal view of time* that emphasizes the similarities between instants of time and the possible worlds in modal logic. Possible worlds are ways the world might have been that collectively form what is known as *logical space*. Whereas the occupants of physical space are material objects that bear the location relation L to metaphysically basic spatial points (substantivalism), or serve as the relata of other material objects (relationism), logical space is populated by sentences or propositions that take truth-values at worlds. Modal views of time treat instants in a smiler manner, as *ways the present was, is, or will be*. In terms of George Myro's (1986a; 1986b) true-at operator |, the time series would

case it would not form a manifold in this narrow mathematical sense.

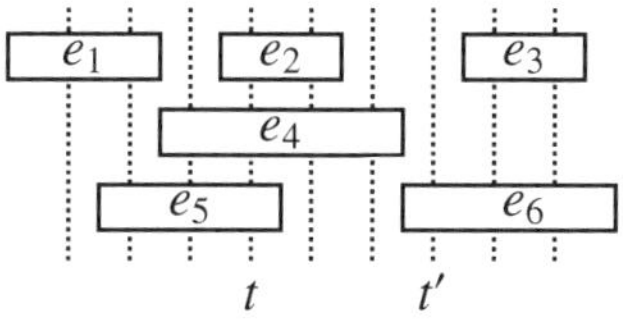

Figure 1.2. Relationism about time

thus get described in terms of claims of the form $t \mid \varphi$, which asserts that the sentence φ is true at the time t. The time series would form a logical space that is occupied by sentences or propositions, rather than a geometric space that is populated by material objects or events.

This gives us three candidate accounts of the nature of time: the temporal substantivalists' claim that time is a manifold of metaphysically basic points, the account of time as a relational structure advocated by relationists, and the modal view that time is a type of logical space. As with other metaphysical theses, these views can be assessed in terms of two types of commitment. The *ontological* commitments of a theory consists of the types of objects whose existence it assumes; its *ideological* commitment comprises all the notions that it takes as conceptually primitive. It is often possible to exchange one type of commitment for the other and we can think of competing metaphysical theories of a subject matter as different ways of making this trade-off.

In our case, spatial views of time are usually spelled out in terms of an extensional logic with appropriately chosen temporal predicates and relations. Modal views of time, on the other hand, are more naturally cast in terms of a tense logic with propositional tense operators P ('it was the case that') and F ('it will be the case that') that behave like the possibility operator $\Diamond$ in modal logic. Relationism about time and modal views thus both propose to trade the metaphysically basic time points of temporal substantivalism for a larger ideological commitment. What they disagree about is what additional primitive notions are needed to compensate for the loss of the substantivalists' ontology. Relationists help themselves to more temporal relations while modal views postulate conceptually primitive tense operators.

Theories with fewer ontological and ideological commitments are usually thought to be preferable, but that alone does not tell us how much is too much, and how little is too little. Parsimony considerations can only decide amongst rival theories if there is some agreement on

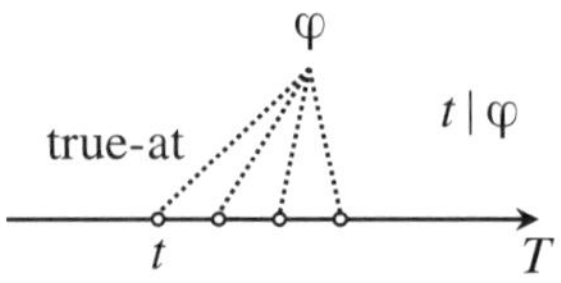

Figure 1.3. Time as a logical space

what theoretical work a theory of time needs to perform. Given a set of problems that need solving, we can say that a theory of time is ideologically or ontologically extravagant if it makes more assumptions than are needed to get the job done, or that it takes parsimony too far if it fails to endow time with enough structure to perform its theoretical role.

As I see it, there are three issues that a theory of time needs to address. First and foremost, it must solve the *Problem of Change*,[4] which is the task of reconciling the possibility of change with the Indiscernibility of Identicals. An object undergoes change just in case it possesses some property K at one time that it lacks at another. Suppose that 'a' is a name for the object while it had K, and 'b' a name for the object after it had lost this property. Then the object experiences change if and only if both of the following are true:

$$a \text{ is } K \qquad b \text{ is not } K$$

On the other hand, the principle of the Indiscernibility of Identicals tells us that, whenever two names pick out the same object, their joint referent either has a given property or lacks it:

$$\text{If } a = b \text{ then, for all properties } X,\ a \text{ is } X \text{ iff } b \text{ is } X$$

Since a is K and b is not K, it follows that $a \neq b$. It cannot be the same object that is first K and then not K; change is logically impossible.

It is perhaps difficult to take this problem seriously at first because its solution seems to be entirely obvious. We ignored the fact that a changing object would possess and lack the property *at different times*. As long as $t \neq t'$, both 'a is K at time t' and 'b is not K at time t'' are compatible with $a = b$. Our formulation of the Indiscernibility of

[4]See Lewis (1986c: 202), Haslanger (1989a;b), and Hinchliff (1996).

Identicals is similarly incomplete. While no object can both possess and lack a property simultaneously, it can do so at different times:

If $a = b$ then, for all properties K and times t, a is K at t iff b is K at t

Once the time dependence of property-possession is explicitly acknowledged, the apparent contradiction vanishes. As G. H. von Wright nicely puts it: 'Time is man's escape from contradiction' (1968: 21).

This much everybody agrees on. Opinions begin to diverge at the next step, when it comes to explaining what it is for an object to possess a property *at* a time. That is where the Problem of Change becomes genuinely problematic. All that we got so far is a blueprint for a solution that is cast in terms of unanalysed expressions of the form:

a is K at time t

The challenge is to give a plausible account of the logical form of such time-relative property attributions. Different theories of the nature of time answer this question in different ways, and the ontological costs of solving the Problem of Change often outweigh a theory's other benefits.

A second constraint is that our theory of time must mesh with the role time plays in our best physical theories of motion. Whatever we may say about the nature of time independently of experience, temporal relations are also the subject of often quite sophisticated measurements. Advocates of spatial views of time like to emphasize this requirement because they believe it speaks in favour of their account. The theory of relativity is said to show that time and space are mere facets of one unified spacetime. Rather than a debate between relationists about time and temporal substantivalism, we would thus get a debate between relationists and substantivalists about spacetime. Proponents of modal accounts need to work a little bit harder to reconcile their view with our best physics, but I shall argue that this challenge can be met.

The third problem is that our metaphysical theory must make appropriate contact with our ordinary way of talking about time. Spatial views like to draw attention to the pervasive use of spatial metaphors in temporal discourse, while modal views of time play up the fact that, in many natural languages, verbs are tensed. On its own, neither argument seems decisive. Moreover, our language evolved in response to evolutionary pressures and communicative challenges that are too far removed from the concerns of metaphysics to be taken all that seriously.

But the folk theory of time that is fossilized in the structure of our language does help to single out what the philosophy of time is talking about. We want an account of what *we* call time and we cannot reject too much of the folk theory without changing the topic.

A familiar phenomenon in empirical enquiry is the underdetermination of theory by observational data. There are always many mutually incompatible empirical theories that can account for all the available evidence. A similar issue arises in metaphysics, where one might worry that some questions are underdetermined by the available problems, but I do not think that this is the case in the philosophy of time. Taken together, the Problem of Change, the role of time in physical theories, and our folk theory of time suffice to settle the issue, and I hope to show that they do so in favour of a modal theory of time.

2

RELATIONISM ABOUT TIME

TEMPORAL substantivalism and relationism about time both treat instants like places, but disagree about whether this requires metaphysically basic time points. Since theories with smaller ontological commitment are usually thought to be preferable, let me begin by considering relationism about time. There are two main versions of this view. One variant treats times as classes of temporally ordered events, the other constructs instants out of the temporal parts of material objects.

2.1 Times as classes of events

Relationism about space tries to avoid ontological commitment to metaphysically basic spatial points by abstracting locations from material objects that stand in spatial relations to one another. How precisely this abstraction is supposed to work is a nice question, and one that relationists often have too little to say about, but at least it is clear where we would begin. This is not true for relationism about time.

While material objects do enter into spatial relations, the problem is that they do not seem to bear any temporal relations to one another. Consider the following two lists of expressions:

Spatial Case	Temporal Case
... is in the same place as ...	... while ...
... is in front of ...	... before ...
... is behind ...	... after...

The items on the left pick out spatial relations, but their counterparts on the right do not pick out temporal relations between material objects. It makes sense to say 'John *is in the same place as* Jack', but 'John *while* Jack' is simply ungrammatical. To get a sentence out of 'while', we need to complement it with two sentences, as in 'John had a martini *while* Jack fixed the car'. Like the other items in the right-hand column, 'while' is a sentence connective, rather than a binary predicate, and thus of the wrong syntactic category to serve as the building block of a relationism about time. One could of course define temporal relations in terms of temporal sentence connectives. For example, we might stipulate that *Bab* holds if and only if *a* exists before *b* exists, but this does not get the relationist around the difficulty that there do not seem to be any fundamental temporal relations between material objects.

There are two places where adjustments could be made. One possibility is to accept the items in the above list for what they are, binary tense operators, which would lead us to a modal theory of time. The other option is to find entities other than material objects to serve as the relata of temporal relations. An obvious candidate for this role are *events*. It does make sense to say that the Renaissance *preceded* the Reformation, that World War II *followed* World War I, and perhaps we can even say that the English Civil War *overlapped* the Thirty Years' War. Most relationists about time therefore try to construct instants from temporally ordered events, but they tend to disagree amongst themselves about how precisely this abstraction is supposed to proceed.

Alfred North Whitehead defends an account of instants of time that is inspired by the construction of real numbers as nested sequences of intervals of rational numbers. He identifies times with so-called *abstractive classes*, which are sets of events such that:[5]

(i) of any two events in the set one temporally extends over the other.

(ii) there is no event in the set that is temporally extended over by every other event in the set.

An event occurs at such a time just in case it is an element of that abstractive class. One noteworthy feature of this proposal is that the existence of times is not guaranteed by the occurrence of events alone. If an abstractive class contained a shortest event then clause (i) would entail that this event is extended over by every other element of the set, contradicting clause (ii). Every abstractive class must therefore contain an infinite sequence of shorter and shorter events, which means that there are no instants in worlds with finitely many events. Whitehead's times also cannot contain any instantaneous events because every instantaneous event is a shortest event.[6] Unless finite sets of events and

[5]Whitehead (1919; 1920: 86; 1925: 104). In his later writings, the notion of an abstractive class is superseded by that of an *extensive connection* (1929: pt. IV, ch. II). Hammerschmidt (1947) gives a useful survey of Whitehead's philosophy of time.

[6]Whitehead would probably not regard this as a problem because he wants to argue against the view that nature consists of nothing but the distribution of matter at extensionless time points. He objects that such a view would leave no room for the notion of velocity and other states of change (1925: 2). See also (1919: 44–6) and (1938: 207).

instantaneous events can be ruled out for other reasons, there could thus be events that occur without occurring *at* any time.

Bertrand Russell defends a rival version of event-relationism that does not run into these difficulties. His proposal is that instants of time are maximal sets of pairwise overlapping events:

> An instant is ... a group of events having the following two properties: (1) Any two members of the group overlap in time; i.e. neither is wholly before the other. (2) No event outside the group overlaps with all of them. We then define one instant as earlier than another if there is some event at the one instant which is earlier than (i.e. wholly precedes) some event at the other. ... an event is 'at' an instant when it is a member of the class. (Russell 1936: 216)

This is the relationist view depicted in Fig. 1.2 above. If we use $\circ$ for the relation of temporal overlap and $\ll$ for the relation of complete temporal precedence then Russell's definitions are:

$$
\begin{array}{lll}
t \text{ is a } \textit{time} & \text{iff} & \forall e_1(e_1 \in t \leftrightarrow \forall e_2(e_2 \in t \rightarrow e_1 \circ e_2)) \\
t_1 < t_2 & \text{iff} & \exists e_1 \exists e_2(e_2 \in t_2 \wedge e_2 \in t_2 \wedge e_1 \ll e_2) \\
e \textit{ occurs} \text{ at } t & \text{iff} & e \in t
\end{array}
$$

Every set of events delivers enough instants to ensure that every event occurs at some Russellian instant, including finite sets of events and those containing instantaneous events.[7]

For both proposals, it is a non-trivial matter how the structural properties of the time series depend on what events happen to occur. For example, suppose we adopt the following three axioms for $\circ$ and $\ll$:

$$
\begin{array}{ll}
\text{R1} & \forall e \neg e \ll e \\
\text{R2} & \forall e_1 \forall e_2 \forall e_3 \forall e_4((e_1 \ll e_2 \wedge e_2 \circ e_3 \wedge e_3 \ll e_4) \rightarrow e_1 \ll e_4) \\
\text{R3} & \forall e_1 \forall e_2(e_1 \circ e_2 \leftrightarrow \neg(e_1 \ll e_2 \vee e_2 \ll e_1))
\end{array}
$$

R1 prohibits events from preceding themselves and R3 allows us to define overlap in terms of precedence by stipulating that two events overlap if and only if neither wholly precedes the other. R2 can be thought of

[7]The proof of the existence of Russellian instants requires the axiom of choice. See also Russell (1915: ch. IV; 1927: 299–302; 1948: pt. IV, ch. V), Kamp (1979), Thomason (1984; 1989), and Anderson (1989). Like Whitehead, Russell is not concerned with reducing ontological commitment. For him, the motivation for event-relationism is primarily epistemological. The main task is to explain how times can be constructed out of immediate sense awareness (1915: 116). Similar views are held by Walker, who claims that 'un instant n'est pas une expérience de base, physique or psychologique, mais est un concept dérivé d'expériences ayant une certain *durée* temporelle' (1947: 131).

as a generalized version of the transitivity of the complete precedence relation on events.[8] Given these three axioms, Russell's earlier-than relation is then guaranteed to be irreflexive, transitive, and comparable.

Here is a quick proof. Suppose $t < t$. By definition of $<$, there are then events $e_1, e_2 \in t$ such that $e_1 \ll e_2$. Because the two events are in the same time they must overlap, $e_1 \circ e_2$, which entails $\neg e_1 \ll e_2$ with R3. Since this is a contradiction, it follows that no time can be earlier than itself and that $<$ is irreflexive. To show the transitivity of $<$, suppose that $t_1 < t_2$ and $t_2 < t_3$. Then there are events $e_1 \in t_1$, $e_2 \in t_2$, $e_3 \in t_2$, and $e_4 \in t_3$ such that $e_1 \ll e_2$, $e_2 \circ e_3$, and $e_3 \ll e_4$. This entails $e_1 \ll e_4$ by R2, and thus $t_1 < t_3$. If one time is earlier than a second that is earlier than a third then the first time is always earlier than the third. Comparability requires that one of two different times is earlier than the other. Suppose that $t_1 \neq t_2$. Since times are sets of events, two times can only be distinct if one contains an event that is not in the other. Without loss of generality, suppose that $e_1 \in t_1$ but $\neg e_1 \in t_2$. Then there must be an $e_2 \in t_2$ such that $\neg e_1 \circ e_2$ because otherwise e_1 would be in t_2. Thus $e_1 \ll e_2 \vee e_2 \ll e_1$ by R3. Hence either $t_1 < t_2$ because e_1 precedes e_2, or $t_2 < t_2$ because e_2 precedes e_1.

These are useful results, but it is far from obvious whether any of R1–R3 are conceptual truths about precedence and overlap. For example, if $<$ is both transitive and irreflexive then the time series cannot be *circular*. If the earlier-than relation is transitive then every time in a temporal loop is earlier than itself, contradicting irreflexivity. And if $<$ is comparable then the time series cannot *branch*. Suppose the time series splits into two future-directed branches at some time t. Then the times in one branch are neither earlier nor later than the times in another branch, even though all of them are later than t. Anybody who admits the possibility of branching or circular times would thus have to deny that the conjunction of R1–R3 is necessary. The irreflexivity, transitivity, and comparability would instead be contingent features that depend on what events happen to occur, and on whether the set of all events satisfies R1–R3.

To get a *dense* time series, we need assurance that there is always a third time between any two given ones. To enforce this feature, Russell (1915: ch. 4) proposes the following condition:

[8] This principle is ultimately due to Wiener (1914: 442). It was later rediscovered by Walker (1947: axiom 2) and called 'Walker's postulate' by Whitrow (1980: sec. 4.6).

R4 $\forall e_1 \forall e_2 (e_1 \ll e_2 \rightarrow \exists e_3 \exists e_4 (e_1 \ll e_3 \wedge e_3 \circ e_4 \wedge e_4 \ll e_2))$

One can then show that, whenever $t_1 < t_2$, there is a Russellian instant t_3 such that $t_1 < t_3$ and $t_3 < t_2$. But since R4 makes substantial assumptions about the existence of events, nobody who does not already believe in the denseness of the time series is likely to endorse this principle.

To get a time series that is isomorphic to the real numbers $\mathbb{R}$, the earlier-than relation would also need to be *complete*. Suppose a set of times is bound from above in that no instant in the set is later than some fixed time, which need not itself be a member of the set. Then the time series is complete just in case every such bounded set is possesses a *least upper bound*, which is the earliest time such that no time in the set is later than it. Examples of complete sets are the real numbers, the naturals and the integers. The rational numbers are not complete.

To guarantee a complete time series, it is more convenient to adopt a proposal due to A. G. Walker (1947). His definition of instants is inspired by the construction of real numbers as Dedekind cuts on the rational numbers. Let H ('*H*istory') be the set of all events, equipped with the precedence relation $\ll$ and the overlap relation $\circ$, which are assumed to satisfy R1–R3. Walker then identifies times with separations of H into past, present, and future events. A time is a triple $\langle A, B, C\rangle$ of sets of events A, B, and C such that:

(i) the union of the three sets exhausts history, $H = A \cup B \cup C$;
(ii) every event in A wholly precedes every element of B;
(iii) every element of C overlaps with some event in A and with some event in B; and
(iv) A and B are non-empty, but C may be empty.

An event e occurs at a time $\langle A, B, C\rangle$ if $e \in C$, and one time is earlier than another, $\langle A, B, C\rangle < \langle A', B', C'\rangle$, if A is a proper subset of A'. Equipped with this earlier-than relation, the set of all Walker times is complete.[9]

On the other hand, Walker's proposal cannot deal with instantaneous events any more than Whitehead's abstractive classes. Since an instantaneous event cannot overlap two events unless they overlap one another, clause (iii) prevents such an event from being in the set C of any Walker-time $\langle A, B, C\rangle$. Instantaneous events can be included in either A or B, but

[9] See Walker (1947: thm. 17). A similar proof is given in Whitrow (1980: sec. 4.6). Both contain a mistake that gets corrected in Thomason (1984: prop. 3). Forbes (1993) and Pianesi and Varzi (1996) discuss additional proposals for abstracting times from events.

they cannot occur at any time. The proposal also does not leave room for a time series that is dense without being complete and it would thus be impossible for the time series to look like the rational numbers.

Hans Kamp (1979: 379–80) tries to get a complete time series in a different way, by adding another step to Russell's construction of instants. His strategy is to apply Russell's abstraction procedure once again to the dense total order of time points we get by adopting R1–R4. This almost yields the desired result, but not quite. There can be two sequences of times, one ascending and the other descending, that intuitively converge to the same time, yet Kamp's way of enforcing completeness puts *two* time points between them. Suppose the set of all events H looks like the set of all non-empty open intervals of real numbers. Let (x, y) be the open interval that contains all numbers that are strictly larger than x and strictly smaller than y. If we can identify events with open intervals in this way, then the following Russellian instants are both candidates for time zero:

$$\{(x, y) : x < 0 \wedge y > 0\} \cup \{(0, z) : z > 0\}$$
$$\{(x, y) : x < 0 \wedge y > 0\} \cup \{(z, 0) : z < 0\}$$

Where we expected one time point, we get two.[10] Kamp presents this as a peculiarity of the approach, but it actually points to a more general problem for event-relationism. So far, we worried that there might be too few instants to form a dense and complete time series. It turns out that event-relationism also suffers from the opposite problem of running the risk of yielding too many times.

To a mathematician, this will not come as a surprise. The constructions of Russell and Whitehead can be regarded as precursors of what is known as a *Stone space*. Suppose we take the set of all events H and equip it with the binary parthood relation that smaller events bear to larger events of which they are episodes. (The Battle of the Somme is part of World War I in this sense.) By making suitable assumptions about the set of all events H, we can define the *join* of two events as the smallest event having both of them as parts, and their *meet* as the largest event common to both of them. If we do this right then H forms a Boolean algebra. To get this result, we need to be fairly generous about

[10]This problem was first noted by de Laguna (1922a;b). Whitehead tries to address this problem in (1929), but with mixed success. For discussion, see Grünbaum (1953: sec. 4) and especially Bostock (2010: sec. 5).

what we count as events, but otherwise this is just like the set of events that Russell and Whitehead start out with.

Given such a set of events, the associated Stone space then consists of all the *ultrafilters* of H. Ultrafilters are similar to Russellian instants but they need to satisfy slightly more stringent conditions. In particular, the join of any two elements of an ultrafilter must also be an element, and ultrafilters must be upwards closed in that anything that one of its elements is part of is itself an element of the ultrafilter. In light of these parallels, one might therefore suggest that Russell and Whitehead were really groping for the idea that the time series is the Stone space of the Boolean algebra of events.[11]

This places event-relationism in a broader theoretical framework and it also helps to identify the root of the problems encountered by Kamp. If H looks like the set of all non-empty open intervals of real numbers then the associated Stone space not only contains *more* than continuum many ultrafilters, it is also *completely disconnected* when equipped with its natural topology.[12] To avoid these problems, event-relationists might have to follow Peter Roeper (1997) and consider suitable chosen sets of ultrafilters as instants, thus reducing the cardinality of the time series to the desired level.

However, there is one key point likely to remain unaffected by the way event-relationists resolve these various technical difficulties. On any plausible way of abstracting times from events, most structural features of the earlier-than relation reflect *contingent* facts about what sort of events happen to occur, rather than necessary properties of the time series. If there are enough events temporally arranged in the right sort of way then the resulting time series might look like the real numbers, but it could also take any number of other shapes. I do not think that, in the final analysis, event-relationism is a tenable view about the nature of time, but I do believe that it is right about this particular issue. The structure of the time series is indeed a contingent matter.

[11]This is the position taken by Mormann (1998; 2005). An accessible introduction to Stone's theorem is in Davey and Priestley (1990: chs. 7–10); see also Johnstone (1982).

[12]A subset of a topological space is connected if and only if it cannot be represented as the disjoint union of two or more non-empty open subsets. In a completely disconnected space, the only connected subsets are singleton sets.

2.2 Are events metaphysically basic?

Some philosophers take event-relationism one step further and propose to identify temporal relations with certain causal relations. There is an interesting literature on such *causal theories of time*, but this is not something I shall discuss here.[13] I want to argue that there are general reasons for rejecting *any* event-relationism about time. If that is right, then the further question of whether temporal relations on events can be analysed in terms of causal relations is irrelevant.

Given that they serve as the building blocks of their account of time, event-relationists need to tell us something about what *events* are. At first sight, this might not seem to be a difficult problem, for this question has received a great deal of attention over the last few decades and there are a number of well-developed accounts of eventhood to chose from:

1. W. V. Quine (1960: 171) and E. J. Lemmon (1966) identify events with spatiotemporal regions.
2. David Lewis (1986b) takes events to be properties of spatiotemporal regions, which allows for finer distinctions than the Quine–Lemmon view.
3. Jaegwon Kim (1993) treats events as compounds $[x, K, t]$ of a substance x, a property K, and a time t. An event exists (occurs) just in case x is K at t.
4. Roderick Chisholm (1976: ch. IV.6) claims that events are special types of states of affairs. Very roughly, his view is that an event is a state of affairs that is not a proposition and is 'concretized' at some place and time. The notion of concretization is spelled out in terms of objects' having certain properties at certain times.
5. Johan van Benthem (1983: ch. I.5) identifies events with pairs $\langle I, \varphi\rangle$ of a sentence φ that is true throughout the time interval I.

This might seem to be a rather large number of competing views about a fundamental ontological category, but all of these proposals agree on one key point. They all endorse Peter Strawson's (1959: 46) view that material objects enjoy ontological priority over events, and that we have

[13] The view itself can be traced back at least as far as Leibniz (1863). See Reichenbach (1924b), van Fraassen (1970: sec. II.3), and Carrier (2003). Prominent advocates in the twentieth century include Carnap (1925), Reichenbach (1924a: §6; 1928: §21; 1951: ch. 9; 1956: §3), and Grünbaum (1963b; 1967). For an assessment of Grünbaum's view, see Lacey (1968); a discussion of Carnap's view is in Grünbaum (1963a). Tooley (1997: ch. 9) considers a causal theory of time in a substantivalist context. See also van Fraassen (1970: ch. VI), Sklar (1974: sec. IV.E), and Le Poidevin (2003: ch. 12).

settled what events occur once we have specified what properties all objects possess at all times. All of them regard events as metaphysically complex entities that are made up of material objects, properties, and times. Their disagreement merely concerns the *number* of events that are to be constructed from these ingredients. But this also means that event-relationists cannot accept any of these accounts of eventhood because they all assume the very thing relationists want to abstract from events: *times*. If such a Strawsonian account were correct then times would be constituent parts of events, rather than the other way around.

The problem is not that we would explain one temporal notion in terms of another. Such conceptual circularity seems unavoidable. Just as one cannot explain modal notions in non-modal terms, it seems futile to hope for an analysis of temporal notions that does not take some other temporal notion as primitive. The issue we are facing here, though, is one of ontological circularity. While relationists claim that times are abstractions from events, Strawsonian accounts of eventhood suppose that events are composed, amongst other things, of times. If events are made up of times then times cannot also be made up of events.

Nor is this meant as an objection to Strawsonian accounts of eventhood. Most contemporary philosophers are interested in events because they believe them to be the relata of causal relations, and some of them also accept Donald Davidson's view that events are needed in the semantics of certain adverbial constructions.[14] The debate is about which of the various proposals best accounts for the role of events in these two applications. None of the standard accounts of eventhood was ever meant to support event-relationism about time, and none of them does.

To treat times as classes of temporally ordered events, we need a different account of what events are. If we assume, as all Strawsonian theories of eventhood do, that events are metaphysically complex entities then we run into a problem that was first raised by Russell, a few years before he became an event-relationist himself:

> The relational theory may seem, at first sight, simpler than the absolute theory, but in its application a great difficulty arises from the absence of any such class of entities as the events which it requires ... Whatever can ... recur or persist, is not an event; but it is difficult to find anything logically incapable of recurrence or persistence, except by including temporal position in the definition.
>
> (Russell 1901: 295)

[14] See Davidson (1980a;c), Lewis (1986a;b), Lombard (1986), Bennett (1988), Casati and Varzi (1996), and Higginbotham et al. (2000).

There can be numerous events, such as the repeated oscillations of an electron, that only differ in the time of their occurrence, but are qualitatively indistinguishable otherwise. If events are incapable of recurring then these are different events, rather than the same event repeating itself. Russell's question is how these events differ from one another.

An obvious answer would be to say that such events differ by the time of their occurrence, but this answer is inadmissible if events are metaphysically complex entities. In that case, their construction must be completed, and their identity conditions settled, before we can talk about events entering into temporal or other relations to one another. Metaphysically complex events could inherit their temporal relations from the temporal features of their constituents, but that is not a view event-relationists can endorse without undermining their central contention that events are the basic temporal entities. That role would instead be played by the temporally ordered constituents of events. If events are incapable of recurring then they cannot be constructed out of something else *and* be the primary occupants of the time series.

One could try to reject the assumption that events cannot recur, but that is unlikely to help event-relationism. To admit the possibility of recurrence is to treat events on the model of properties, which can have multiple instantiations. This might be a viable alternative to the standard accounts of eventhood listed above, but it immediately leads us to the question of what instantiates such event-properties. The only plausible answer seems to be Richard Montague's (1969) view that these event properties are instantiated by times. In that case, the account of the event-tokens that enter into temporal relations—the possessions of event-properties by *times*—would again assume the very entities that relationists are trying to abstract from them.

As one can tell from the quote on p. 9, Russell later endorsed the very event-relationism he deemed beyond repair in 1901. I do not know why he changed his mind, but one obvious strategy for avoiding these difficulties is to regard events as metaphysically basic entities, which is also the view held by Whitehead, Russell's main collaborator during this period. If events are fundamental entities then we do not need to provide identity conditions for them. We could start out with a collection of events that stand in temporal relations to one another. That there is one wiggling of an electron followed by another wiggling would be a brute fact about the world. After we have applied Russell's abstraction procedure to this case, we could distinguish these intrinsically indistinguishable events by their time of occurrence. It would be necessarily true that successive wigglings of the electron occur at different times, but this would be a consequence of the abstraction of times, not a report on the constituent components of events.

This takes care of the temporal relations between events; the problem is that it leaves us without an account of how material objects relate to time. If events are the primary occupants of the time series then material objects relate to time only indirectly, by *participating* in events that stand in temporal relations to one another. The challenge is to explain what such participation consists in. Strawsonian accounts of eventhood usually give us an account of participation for free. On Kim's proposal, for instance, we could say that an object a participates in the event $[b, K, t]$ if and only if $a = b$. Similarly, the Quine–Lemmon view allows us to say that an object participates in an event if it exists when the event occurs and is located where it happens. These options are not available if events are metaphysically basic.

Instead of participation, Whitehead (1925: §53) talks about objects being 'situated' in an event, which he takes to be a conceptually primitive notion. But objects are clearly not 'situated' in events the same way in which, say, my books are situated inside my office. Location is an external relation between material objects and places. Whether or not a book is located in my office depends on more than the intrinsic natures of the book and the office. The office and the book can both exist without the latter being located in the former. Participation, on the other hand, is an internal relation. Which objects participate in an event is determined by its intrinsic nature alone. This does not mean that all the objects that participate in an event are essential participants. René Descartes might have participated in the Thirty Years' War, but he hardly ever left camp to do his share of the raping and pillaging and it seems plausible to say that war would still have taken place without him. Even though Descartes was an inessential participant, any counterfactual Thirty Years' War without him is intrinsically different from the war that actually happened. There cannot be two possible worlds that contain the same event e with exactly the same intrinsic features, but in one world a material object a participates in e, and in the other world it does not participate.

If that is right then a theory of events cannot regard participation as a primitive relation. If a relation is internal then there is something else in virtue of which it obtains. The locations of objects might be primitive, unanalysable facts, but participation is a complex modal connection between the existence of objects and the occurrence of events. While I can remove all books without my office going out of existence, there is no event without the participating objects. If none of the participants had existed then the Thirty Years' War would not have happened. But if the war had not occurred then the participants could still have existed; they could just have participated in some other events. Perhaps it is true that every object must participate in some event at every moment of its existence, but there is no specific event that must occur for a particular object to exist (unless we count merely existing as an event).

I want to argue that this asymmetric dependence between the occurrence of events and the existence of material objects is indistinguishable from the way in which metaphysically complex entities depend on their constituent components. To a first approximation, an event's participants are those material objects without whose existence the event would not have occurred. There are different way of spelling out the sense of 'would' at issue, but the details do not matter much in the end. Suppose we are given some conditional $\Rightarrow$ whose strength is somewhere between the material conditional (very weak) and the strict conditional (very strong). Then a material object a participates in event e *if* the event's occurrence is conditional upon that object's existence:

(A) a does not exist $\Rightarrow$ e does not occur

If there can be non-essential participants, such as Descartes and the Thirty Years' War, then (A) is not necessary for participation, but all I need for my argument is that the condition is sufficient. Now suppose that a would not have existed if some other object b had not existed:

(B) b does not exist $\Rightarrow$ a does not exist

Given suitable background assumptions to ensure the transitivity of $\Rightarrow$ in this case, it follows that b participates in e as well. That is, any object whose existence is necessary for the existence of a participant is itself a participant. This has untenable consequences.

Suppose we follow Gideon Rosen (1992: 86–91) and read $\Rightarrow$ as the counterfactual conditional. Suppose further that it is true that we all descended from Adam and Eve. In that case, none of us would have existed if the two of them had not existed and it follows that Adam and Eve participate in my writing of this sentence. If they had not existed then I would not have existed, and if I had not existed then my writing would not have occurred.[15] Similar problems arise if we spell out participation in terms of a strict conditional. Suppose we adopt Saul Kripke's (1980) views on the necessity of origin, and accept that it is an essential property of me that I descended from the parents that I did in fact descend from. Then I would not have existed if my parents had not existed and they would not have existed if their parents had not existed. It follows that also my dead grandparents participate in my writing.

Rosen suggests that we add the requirement that participants must exist *when* the event occurred. On his view, Adam and Eve and my grandparents do not participate in my writing because they no longer exist when it takes place. This might yield a more plausible account of participation, but we cannot adopt this solution if the primary temporal entities are metaphysically basic events. The proposed fix appeals to the notion of an object's existing *at a time*, which is part of what the notion of participation was supposed to provide us with. The plan was to say, after we had secured an account of participation, that a material object exists at a time if and only if it suitably participates in an event that occurs then. If we already need an account of *existence at a time* to spell out the notion of participation then our relationist account of time, events, and material objects never gets off the ground.

The only way of avoiding the problem is by denying that there could be an object *b* that is both wholly distinct from *a* and necessary for its existence. But this would require a reading of $\Rightarrow$ in (B) that is so strong that at best the parts of an object are necessary for its existence. Since we need to use the same conditional in (A), we are bound to end up with an account on which only the constituent parts of events could qualify as participants. We either get an account of participation that is implausibly restrictive, by ruling out all participation, or one that contradicts the assumption that events are metaphysically basic entities that do not have any material objects as constituent components. Any plausible account of participation thus collapses into an account of constitution.

This shows that events cannot be the primary occupants of the time series *and* be metaphysically simple. Unless we regard events as metaphysically complex, we have nothing plausible to say about how ordinary material objects relate to time. And if we have no account of the relation between material objects and time then we also have no ac-

[15] In general, the counterfactual conditional is not transitive, but we only need one counterexample in which the conditions are such as to ensure transitivity in that specific case.

count of what it is for an object to have a property at a time, and thus no solution to the Problem of Change. Since we have already established, with the help of Russell's argument, that events also cannot be the primary occupants of the time series *and* be metaphysically complex, it follows that events are not the primary occupants of the time series. Event-relationism about time is untenable.

2.3 Temporal-parts relationism

According to our ordinary way of talking, only events have temporal parts or stages. For example, the Defenestration of Prague and the Peace of Westphalia are parts of the Thirty Years' War. On the other hand, we do not usually say that Descartes' military career during this war or his sojourn at the Swedish court are parts of *him*. At best, they are parts of an event, his life, in which he participated. Yet a fair number of contemporary philosophers think that this aspect of our pre-theoretic conception of material objects is radically mistaken. They believe that the right way of thinking about objects is in terms of what Peter Strawson dismissively calls *process-things*: 'four-dimensional objects ... of which each of the successive temporal parts is three-dimensional, is, as it were, the thing taken at successive stages of its history from the beginning to the end' (1959: 56–7). Just as material objects extend through space by possessing spatial parts at different locations, material objects are said to persist by possessing different temporal parts at different times. Following Lewis (1986c: 202), the acceptance of temporal parts is often called *perdurantism* and the opposing position *endurantism*.[16]

Perdurantists could argue that temporal parts are the primary relata of temporal relations, and that times are to be identified with appropriately chosen classes, or mereological sums, of temporal parts. Both Gustav Bergmann (1960) and Fred Dretske (1961) think of relationism about time in these terms. Once we have abstracted times from temporally ordered temporal parts we could then identify temporal parts by the time of their location and say that an object a exists at time t just in case it has a temporal part a-at-t. The Problem of Change would get solved in a similar manner. If the object a has a property K at time t that

[16] Perdurantism can be distinguished further. *Worm theories* hold that material objects are four-dimensional entities ('spacetime worms') of which only a three-dimensional slice is present at each time. *Stage theories* claim that all of a material object is present at the one time at which it exists, and that it merely has counterparts at other times. Since this distinction does not matter for current purposes, let me ignore it here. See Sider (2001: ch. 5) for a more detailed discussion.

it lacks at time t' then that just means that there are two temporal parts a-at-t and a-at-t' such that the former is K and the latter is not:

$$K\,a\text{-at-}t \qquad \neg K\,a\text{-at-}t'$$

As long as $t \neq t'$, the two temporal parts a-at-t and a-at-t' are different objects, so there is no conflict with the Indiscernibility of Identicals.[17]

Before they can reap these theoretical benefits, temporal-parts relationists would have to provide us with an account of what it is for one object to be a temporal part of another. Most authors follow Judith Jarvis Thomson (1983), who suggests that an object x is a temporal part of an object y if and only if:[18]

(i) Both x and y exist during some time interval I.
(ii) Object x has no spatial part that exists at any time outside I.
(iii) Objects x and y are located at exactly the same places during I.

Instantaneous temporal parts correspond to time intervals that consists of single instants. I think this is a plausible view of what temporal parts would be like, if there were any, but it is not an account that temporal-parts relationists about time can adopt. Like the Strawsonian accounts of eventhood discussed earlier, this definition appeals to the very entities, *times*, that relationists want to construct out of temporal parts. On Thomson's account, temporal mereology recapitulates temporal topology, rather than the other way around.

A more suitable account of temporal parts is proposed in §52 of Rudolf Carnap's *Introduction to Symbolic Logic* (1958), which bears the slightly misleading title 'Axiom Systems of Biology'. Instead of explicating temporal parthood in terms of ordinary parthood and existence at a time, Carnap helps himself to a generalized parthood relation $<$ and a 'before' relation B on such parts. According to Carnap, an object x is a momentary part of an object y if and only if $x < y$ and no part of x bears B to any other part of x. Instantaneous temporal parts are defined as maximal momentary parts; that is, as momentary parts that are not proper parts of any other momentary part.

One might object that this proposal only succeeds because it assumes from the outset that all controversial issues have been settled

[17] Mellor (1998: sec. 8.4) objects that this is not a solution to the Problem of Change, but a concession that no object can ever lose or acquire any properties. Hawley (2001: 93) disagrees; see also Heller (1992) and Lombard (1994).

[18] See, e.g. Lewis (1983b: 76), Simons (1987: 132), Heller (1990: 12), Sider (2001: 59), and Hawley (2001). Unlike these authors, Thomson *rejects* perdurantism.

in its favour. For one, Carnap's generalized parthood relation seems to gloss over the significant differences between spatial and temporal parts that an account of temporal parts is supposed to illuminate. Given a time interval *I*, Thomson's definition would count the temporal cross-section *a*-during-*I* as a temporal part of the material object *a*. But if *R* is a region of space then the spatial cross-section *a*-in-*R* need not qualify as an ordinary spatial part of *a*. As Jeremy Butterfield (1985: 37) notes, material objects usually take their parts with them when they move through space and they can also acquire or lose parts. The material contents of the same spatial region *R* can therefore be part of *a* at one time and fail to be part of it at another. As far as ordinary spatial parts are concerned, it only makes sense to say that the contents of a region are part of the object *at* some particular time. Given a material object *a*, a suitably well-behaved region of space *R*, and a time *t*, the most we can say about the relation between parts and spatial regions is that the overlap of the object *a* with the region *R* at time *t* is a part of the object *a* at that time. Like other spatial relations, ordinary parthood obtains *at* a time.

Similar worries arise for Carnap's other primitive, the 'before' relation *B*. We noted earlier that material objects do not seem to enter into fundamental temporal relations to one another. The only sense we could make of an object *a* being 'before' object *b* was by taking this as shorthand for claiming that *a* exists before *b* exists, or perhaps that *a* has some property *F* before *b* has some property *G*.

(i) Fa before Gb.

(ii) There is a temporal part x of a and there is a temporal part y of b such that Fx, Gy, and Bxy.

Since the primitive notion is the 'before' operator in (i), not the 'before' relation in (ii), any attempt by temporal-parts relationists to offer (ii) as an account of (i) would thus seem to get matters backwards.

But perhaps this is the wrong way of thinking about the proposal. Instead of regarding (ii) as an account of the meanings of 'part' and 'before', one could take it as an attempt at spelling out the deep metaphysics behind (i). The conceptual primitives that make up the ideological commitments of our metaphysical theories need not be definitionally primitive or intuitively graspable to be acceptable as primitives. Temporal-parts relationists could reject the demand that they spell out the relations $<$ and B in terms of an already understood, pre-theoretic vocabulary. Instead, they might argue that these notions are implicitly defined by our best metaphysical theories, similar to the way theoretical terms like 'force' are defined by our best physical theories. While the theoretical notion of force has some connection to our everyday concept, it is not an argument against a physical theory that it also uses 'force' in novel ways. Temporal-parts relationists could adopt a similar stance with regard to their innovative use of 'part'.

Something like this is probably the most plausible version of relationism about time. My only reason for rejecting it is that it compares poorly with modal accounts. Suppose we ask the temporal-parts relationist *how many* temporal parts there are. For the view to work as advertised, there would need to be enough temporal parts to allow us to account for every truth of type (i) in terms of a truth of type (ii). On the other hand, there seems to be no justification for postulating more temporal parts than are needed for this purpose. In the final analysis, we thus get nothing but a fancy way of accounting for claims of type (i). By contrast, the modal account of time I favour does not try to analyse tense operators in terms of anything else and terminates its chain of analysis with claims like (i). For the higher ontological costs of temporal-parts relationism to be justified, there would have to be serious obstacles to formulating a theory of time in terms of primitive tense operators, and I shall argue that such modal accounts are in fact unproblematic. If that is correct then temporal-parts relationism is ontologically extravagant.

3

TEMPORAL SUBSTANTIVALISM

Relationism about time runs into difficulties because it tries to construct times from temporal relations. By taking times to be metaphysically basic entities, temporal substantivalism avoids such problems from the outset. This is an advantage, but temporal substantivalism fails to live up to the promise of treating times like places. Temporal substantivalism is either unable to solve the Problem of Change, or it ends up being little more than an ontologically costly variant of a modal account of time.

3.1 Times and properties

At first sight, temporal substantivalism looks like an exact counterpart of spatial substantivalism. On one side, we have the manifold $\mathcal{M}$ of metaphysically basic spatial points to which material objects bear the location relation L; on the the other side, we have the temporal manifold $\mathcal{T}$ of metaphysically basic temporal points to which material objects bear the exists-at relation E. But there is an important residual difference between space and time that quickly destroys this apparent symmetry. The underlying problem is nicely summed up by Michael Dummett:

> A spatial reference . . . is most naturally construed as a predicate which is true of a given object at a given time: thus 'There are snakes in England' is of the form 'There are snakes which are in England' . . . A temporal reference . . . qualifies the entire sentence adverbially: 'John is ill today' plainly cannot be interpreted as of the form 'John is ill and John is today'. (Dummett 1981: 389)

A spatial substantivalist would regard 'There are snakes in England' as an existentially quantified conjunction of the form $\exists x (Sx \wedge Lxe)$. The first conjunct attributes the intrinsic property of being a snake, Sx, and the second conjunct ascribes spatial location in England, Lxe. The difference is that a temporal substantivalist cannot likewise isolate the temporal location in 'John is ill on Monday', by taking it to claim that John is ill and exists on Monday. If $Ij \wedge Ejm$ were the correct way of regimenting this claim then it would be impossible for John's health to improve. His being healthy on Wednesday would require the truth of

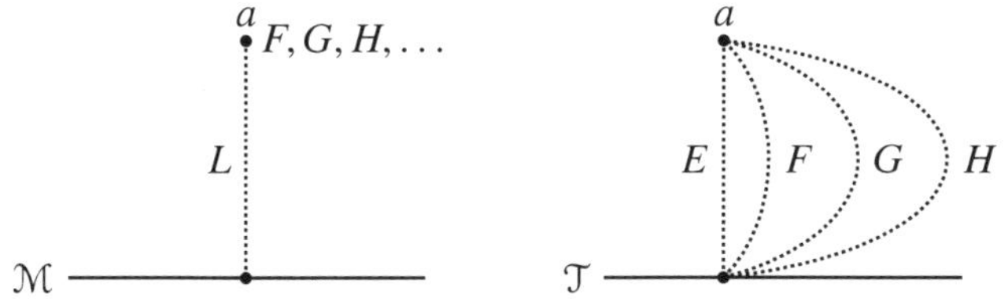

Figure 3.1. Spatial and temporal substantivalism

$\neg Ij \wedge Ejw$, which would contradict his being ill on Monday. More generally, if $Ka \wedge Eat$ were the logical form of 'a is K at time t' then a cannot acquire or lose K. Change would be logically impossible. If we construe temporal substantivalism as a strict counterpart of spatial substantivalism then we would be unable to solve the Problem of Change.

There is a second version of temporal substantivalism that does better in this respect. Rather than regiment 'John is ill on Monday' in the self-defeating manner just described, Gustav Bergmann (1960: 230) and the early D. H. Mellor (1981a: ch. 7) suggest that it is of the form Ijm, where I is the ill-at *relation* that living creatures bear to times at which they are unwell. There is no conflict between John's being ill on Monday and his being well on Wednesday, which would require the truth of $\neg Ijw$. To say that object a has property K at time t would be to say that a bears the K-relation that to point t the temporal manifold, Kat. The Indiscernibility of Identicals would take the form of the schema:

$$a = b \rightarrow \forall t\,(t \in \mathcal{T} \rightarrow (Kat \leftrightarrow Kbt))$$

More generally, if, at some time t, objects $a_1, \ldots, a_n$ are said to stand in some n-ary relation then this modified version of temporal substantivalism would claim that there is really an $(n + 1)$-ary relation between the objects $a_1, \ldots, a_n$ and the point t on the manifold $\mathcal{T}$.

This provides us with a solution to the Problem of Change, but the resulting account of time is quite unlike the spatial substantivalism that served as its inspiration. In the spatial case, there is ultimately only one relation that material objects bear to the spatial manifold $\mathcal{M}$, namely the spatial location relation L. All other spatial properties, such as the shape of an object, depend on where its parts are located. In addition, there are intrinsic properties $F, G, H, \ldots$, such as charge or mass, that an object possesses independently of its location. Temporal substantivalists might propose the exists-at relation E as the temporal analogue of L, but if they solve Dummett's problem by treating all features of material objects as relations to time points then they can offer us nothing comparable to these location-independent intrinsic properties. This erodes the difference between temporal substantivalism and modal views of time.[19]

If material objects have no intrinsic properties, and if all their features are relations to points on $\mathcal{T}$, then we can rewrite temporal substantivalism as a theory of times as indices at which sentences take truth values. Suppose we are given a claim of the form *Kat* that reports that object a stands in the K-relation to some point t on the temporal manifold $\mathcal{T}$. We then introduce a new monadic predicate K^* and a true-at operator | between time points and sentences. This allows us to rewrite *Kat* as $t \,|\, K^*a$, which asserts that K^*a is *true at* time t. More generally, we introduce an n-ary relation Φ^* for every $(n+1)$-ary relation Φ between objects and time points. The true-at operator | between times and sentences containing only starred predicates is then defined as follows:

$$
\begin{array}{lll}
t \,|\, \Phi^*\tau_1 \ldots \tau_n & \text{iff} & \Phi\tau_1 \ldots \tau_n t \\
t \,|\, \tau_1 = \tau_2 & \text{iff} & \tau_1 = \tau_2 \\
t \,|\, \neg\varphi & \text{iff} & \neg t \,|\, \varphi \\
t \,|\, \varphi \rightarrow \psi & \text{iff} & t \,|\, \varphi \rightarrow t \,|\, \psi \\
t \,|\, \forall\xi\, \varphi & \text{iff} & \forall\xi\, t \,|\, \varphi
\end{array}
$$

The τ_i are either individual constants or variables and ξ is a variable. By the first clause, the exists-at relation E between objects and time points gets re-interpreted as a time-relative existence predicate E^*. This translation manual takes it for granted that all un-starred predicates have exactly one temporal parameter, but this assumption is shared by temporal substantivalists, who believe that all temporal relations between material objects can be reduced to the temporal relations between the time points at which they are located.

There is no comparable transformation of spatial substantivalism, even though some authors have tried to provide one. Nicholas Rescher and James Garson (1968) and G. H. von Wright (1979) propose modal accounts of space that treat places as indices at which sentences take truth values.[20] They are impressed by the fact that 'There are snakes'

[19]Lewis (1986c: 202) complains that this proposal would also mischaracterize some clear examples of intrinsic properties as relational. Other authors argue that our intuitions about intrinsicness do not need be taken all that seriously. See, e.g. Mellor (1998: sec. 8.5), Hawley (1998), Wasserman (2003), and Eddon (2010).

[20]See also Mellor (1981a: ch. 4) and Simons (2006). Cresswell (1996) develops a view that treats all predicates as sentences that take truth values at individuals or sequences of

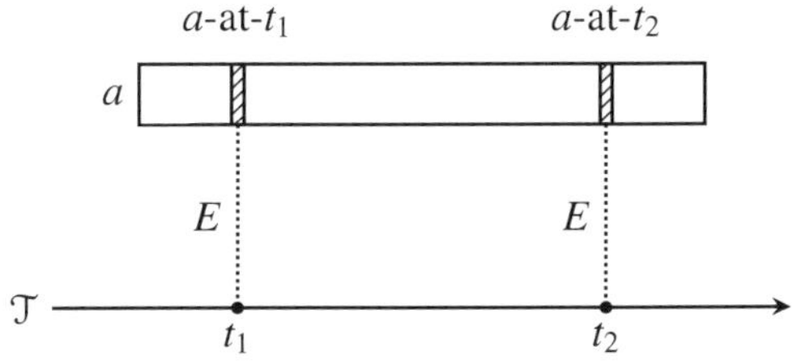

Figure 3.2. Temporal-parts substantivalism

can plausibly be said to be *true at* some places and false at others. Dummett's example 'There are snakes in England', which we earlier formalized as $\exists x\,(Sx \wedge Lxe)$, might thus be re-written as $e \,|\, \exists x\, Sx$. However, this only allows us to rewrite sentences of the special form $Ka \wedge Lap$ as $p \,|\, Ka$. The proposal is unable to accommodate atomic sentences of the form Ka that attribute an intrinsic property K to a material object a. Property attributions are not linked to locations in the same way in which they are linked to times or possible worlds. The apple might be green now and red later, but to say that it is green here either asserts that the apple has a part that is green simpliciter or it is an odd way of asserting the conjunction that the apple is both green and located here. What permits the reinterpretation of temporal substantivalism as a theory of indices is that it requires every property attribution to be located at a time, just as every such attribution must be located in some possible world or other. It thus accepts the central tenet of modal accounts of time; it merely spells out the idea in an ontologically costly manner, by committing itself to the existence of the temporal manifold $\mathcal{T}$.

There is a third version of temporal substantivalism that deals with Dummett's problem by postulating the existence of temporal parts of material objects. Suppose we start out with a temporal manifold $\mathcal{T}$ and then use Thomson's account from p. 21 to define temporal parts in terms of the exists-at relation E that material objects bear to the points on $\mathcal{T}$. At any time t at which it exists, an object a would then possess a temporal part a-at-t that bears the exists-at relation E to that point on the temporal manifold. Instead of re-casting the intrinsic features of material objects as relations to times, such properties would be possessed

individuals. Like the logical spaces discussed by van Fraassen (1970: 100), Cresswell's proposal yields a modal account of ordinary particulars, not a modal account of places.

by the temporal parts of material objects. Just as a poker can have a hot (spatial) part at one end and a cold one at the other, an object can have some temporal parts that possess an intrinsic property and others that lack it. However, such a *temporal-parts substantivalism* still requires every property attribution to be located at some time. The only difference is that the time parameter now gets attached to the temporal part to which the property is being attributed. Suppose we replace the first clause in our translation manual with the following:

$$t \mid \Phi^* \tau_1 \ldots \tau_n \quad \text{iff} \quad \Phi\, \tau_1\text{-at-}t \ldots \tau_n\text{-at-}t$$

In this way, we can also reinterpret the third version of temporal substantivalism as a theory of indices at which sentences take truth values.

3.2 Unemployed time points

The second and third version of temporal substantivalism can be thought of as mere notational variants of a theory of time that treats instants as indices at which sentences take truth values. But both of these proposals differ from other modal accounts of time in that they provide a notion of absolute temporal location that consists in events occurring at particular points on the temporal manifold $\mathcal{T}$. To establish the thesis that temporal substantivalism is ontologically extravagant it remains to be shown that there is no theoretical use for this notion.

The basic idea of the argument is due to G. W. Leibniz (1875–90: vol. VII, 364), who notes in his correspondence with Samuel Clarke that the actual sequence of events is indistinguishable from one in which everything happens a year earlier. Put in slightly more technical terms, Leibniz's observation is that all our physical theories are invariant under temporal translations in which every event in the history of the world gets shifted by the same fixed amount of time. According to temporal substantivalists, such an imaginary shift yields a different state of affairs, either because some object a would bear some relation K to a time t' on $\mathcal{T}$ rather than an earlier time t, or because the temporal part a-at-t' would be K instead of the temporal part a-at-t. Yet no physical process would unfold differently if it were merely shifted to a later time. Leibniz concludes that there is no need for the notion of absolute temporal location that metaphysically basic time points would provide, and hence no reason to postulate the existence of the manifold $\mathcal{T}$.

Time-translation invariance is not an accidental feature of our physical theories. It reflects the homogeneity of time and is also intimately

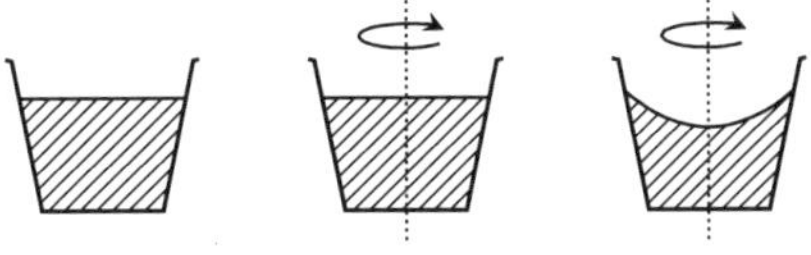

Figure 3.3. Newton's rotating bucket

connected with the conservation of energy. There is a well-known theorem, due to Emmy Noether, that states that every symmetry of a physical theory gives rise to a conserved quantity.[21] This theorem associates time-translation invariance with the conservation of energy. To the extent to which one would expect any plausible physical theory to conserve energy and endorse the homogeneity of time, one would therefore expect it to be time-translation invariant as well.

I agree with Leibniz's conclusion that metaphysically basic time points are a gratuitous addition to our ontology, but his argument contains an important lacuna. It would be hasty to reject temporal substantivalism on the basis of time-translation invariance alone. A similar argument can be given in the spatial case, where the actual arrangement of objects on the spatial manifold is indistinguishable from one in which all objects are shifted by a fixed spatial distance, while keeping all the spatial relations amongst these objects constant. This symmetry reflects the homogeneity of space and is associated by Noether's Theorem with the conservation of momentum. Leibniz takes the invariance under spatial translation as reason for rejecting spatial substantivalism as well, but this argument overlooks the possibility that spatial points can earn their keep in other ways. There are good reasons for postulating metaphysically basic spatial points even though our best scientific theories have no use for the notion of absolute spatial location.

One area where spatial points find employment is in accounting for the observable effects of inertial forces. Isaac Newton (1687: 9) presents the example of a bucket full of water that is suspended from a rope. Initially, the bucket is at rest, and the surface of the water is flat (A). We then begin to rotate the bucket at constant angular velocity around its

[21]Noether (1918). To be more precise, Noether's Theorem is only concerned with differentiable symmetries generated by local actions, and with physical theories that can be formulated in terms of a least-action principle. The details can be found in any textbook on theoretical mechanics, such as Arnol'd (1980: §20) or Goldstein (1980: sec. 12.7).

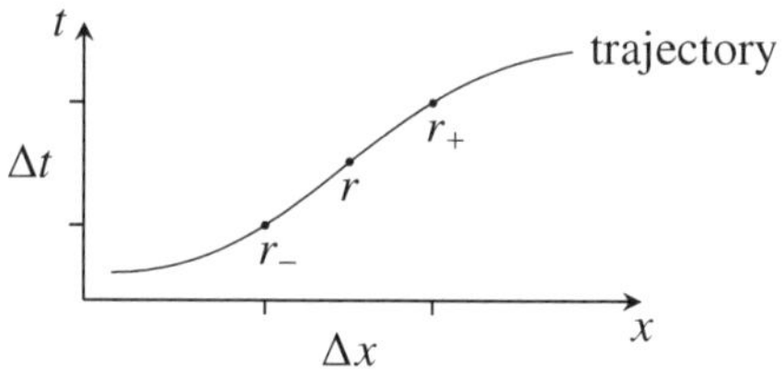

Figure 3.4. Components of acceleration

axis. Due to its inertia, the water in the bucket initially remains at rest (B). After a short while, the rotation of the bucket begins to transmit itself to the water. As the water starts to rotate, the resulting centrifugal force pushes it away from the axis of rotation. The surface of the water becomes more and more concave until the bucket and the water rotate at the same velocity (C). The challenge is to explain why the surface of the water has a different shape in C than it has in A.

Spatial substantivalists have a straightforward answer. In (C), the water is accelerated relative to absolute space and its shape can be explained in terms of the inertial forces (in this case, the centrifugal force) it experiences as a result of this acceleration. There is no such acceleration in (A), so there are no inertial forces to distort the surface of the water. Since there is no relative motion between the water and the bucket in either (A) or (C), it is not clear what relationists about space could say about this case. There might be relative motion between the bucket and external observers, but there is no reason to think that this matters to the water. Rather than rotate the bucket relative to a stationary observer, the same relative motion could be obtained by rotating the observer around the stationary bucket, and we know from experience that running around a bucket does not change the surface of the water.

Inertial forces reveal themselves through acceleration, which is the rate at which an object's velocity changes. Velocity, in turn, is the rate at which an object's location changes. Suppose we simplify matters by considering motion in the x-direction only. Then an object's trajectory can be represented as a curve in the x/t-plane and we can determine the velocity of the object at some point r on the trajectory by performing the following calculation. We pick a point r_- right before r on the trajectory and a point r_+ right after it and then determine the spatial distance Δx and the temporal distance Δt between these two points. The object's average velocity between r_- and r_+ is given by the ratio $\Delta x/\Delta t$ of distance traversed and time elapsed and its (instantaneous) velocity at r is the limit of such average velocities as r_- and r_+ approach r. To determine the acceleration of the object, we perform a similar calculation, with velocities playing the role of distances. Let me pass over the details of this here. The important point is that an account of inertial forces requires an account of quantities like Δx and Δt.

Relationists about space treat spatial coordinates as labels for the positions in the order that spatial relations impose on material objects. One of the distinctive features of spatial relations is that they only obtain between objects that exist at the same time. For one object to be in front of, to the right of, or behind another, both need to exist at the time in question. Since spatial relations always act 'horizontally', parallel to the x-axis, they impose no restriction on how we align the spatial coordinates that we abstract at one time with those that we abstract at another. The distance Δx between an object's locations at different times would thus reflect an arbitrary convention, rather than an objective feature of physical reality. The key contribution of metaphysically basic spatial points is to permit a re-identification of locations over time. While one cannot stand in front of Julius Caesar because he no longer exists, one can stand, say, in front of the region on the spatial manifold $\mathcal{M}$ where he was assassinated. Spatial substantivalists can account for inertial forces because they can take Δx to measure the objective distance between the object's successive locations on $\mathcal{M}$. This provides theoretical employment for metaphysically basic spatial points, even though Newtonian mechanics is invariant under spatial translations.

Whether Newton's rotating bucket does indeed provide a decisive argument in favour of spatial substantivalism is a hotly debated question that we cannot settle here.[22] The point I want to make is that there is no temporal counterpart of this argument. Because spatial relations require the co-existence of their relata, we had to appeal to metaphysically basic spatial points to re-identify locations over time. What is different in the temporal case is that temporal relations do not require the co-location of their relata. We can truly say that Caesar's assassination happened after the crossing of the Rubicon without thereby implying that the two happened at the same place. Temporal substantivalists would say that Δt measures objective temporal distances between points on the manifold $\mathcal{T}$, but their opponents can easily account for such quantities without ac-

[22]For discussion, see Earman (1989) or Pooley (2013).

cepting metaphysically basic time points. Relationists about time only need to postulate enough temporal relations to be able to define quantities like Δt in terms of them. Some relationists might admit to few temporal relations than are needed for this purpose, but there is nothing that prevents them, *qua* relationists, from accepting objective temporal distances between spatially separated events. (Section 7.3 establishes a similar result for modal accounts.) Since all accounts of the nature of time can accommodate quantities like Δx, inertial forces do not even provide *prima facie* reason for favouring temporal substantivalism.

While Newton focuses on the motion of material objects, Hartry Field (1980: ch. 4; 1984: sec. 3) is concerned with the propagation of fields. Unlike earlier theories of mechanics, modern physics does not admit direct contact action of one material object on another; all interactions are instead mediated by force fields, such as the gravitational or the electromagnetic field. Relationists about space might argue that all talk about fields is merely a fancy way of talking about the relative locations of masses and charges. But our physical theories also admit *free* fields that exist without any generating masses or charges. For example, the charges that were responsible for the electromagnetic signals that we are now receiving from some distant stars have gone out of existence billions of years ago. The generating charges are gone, but we still have the fields, which are producing observable effects. Field concludes that we have no choice but to treat field strengths as properties of spatial points. This would provide another way in which metaphysically basic spatial points contribute to physical theory, but also this second argument only lends support to *spatial* substantivalism. Since the manifold $\mathcal{M}$ already provides a medium for fields, there is no work left to be done by metaphysically basic time points.

Even though all our scientific theories are invariant under spatial translations, there can be good reasons for postulating metaphysically basic spatial points. But neither of the two main arguments in favour of spatial substantivalism carries over to the temporal case and Leibniz's dismissal of temporal substantivalism seems to be vindicated in the end. Metaphysically basic time points are dispensable in our best theories of the motion of objects and the propagation of fields.

3.3 The temporal manifold

Metaphysically basic time points are not only redundant, they are also slightly odd. Suppose we inquire about the spatial properties of the

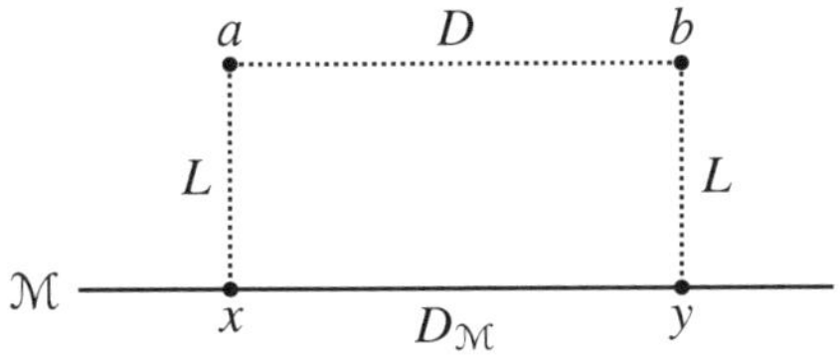

Figure 3.5. Spatial relations

points on the spatial manifold $\mathcal{M}$, by asking *where* they are located. If we are not careful in answering this question, we risk a vicious regress that was already known to ancient philosophers. According to Aristotle, Zeno of Elea complains that 'if everything that exists has a place, place too will have a place, and so on *ad infinitum*' (*Physics* 209^a24).

One option for avoiding this regress is to deny its premise that everything has a place. Spatial substantivalists are offering an analysis of the spatial relations between material objects in terms of the relations between a new class of entities: the points on the manifold $\mathcal{M}$. Suppose we are given two material objects a and b that bear some spatial distance relation D to one another, Dab. According to spatial substantivalism, there are then two points x and y on $\mathcal{M}$ that bear a distance relation $D_{\mathcal{M}}$ to each other and to which a and b bear the spatial location relation L:

$$Dab \text{ iff } \exists x \exists y (x \in \mathcal{M} \wedge y \in \mathcal{M} \wedge Lax \wedge Lby \wedge D_{\mathcal{M}}xy)$$

The relation $D_{\mathcal{M}}$ might have similar structural properties as D but that does not change the fact that it is a different relation. Since the features of $\mathcal{M}$ do not form part of what spatial substantivalists are trying to analyse, they could reject Zeno's question of where spatial points are located as ill-formed. By endorsing the above analysis of D, substantivalists do not commit themselves to the claim that $D_{\mathcal{M}}$ could be analysed in the same way. Indeed, the whole point of their proposal is that $\mathcal{M}$, $D_{\mathcal{M}}$, and L are metaphysically basic, and thus resist further analysis.

A second solution is to begin the regress, by granting his premise that everything has a place, and then to bring it to a swift halt. The only plausible way of extending the location relation L to the elements of the spatial manifold $\mathcal{M}$ itself is by stipulating that every spatial point bears this relation to *itself*, and to itself only. If being located at a certain place is to occupy a position on $\mathcal{M}$ then any point on the manifold ought to be in the same place as any material object located there. Everything would indeed have a place, but this does not lead to an infinite regress because the place of a place would always be the place itself.

Similar issues arise for temporal substantivalism, where one might ask *when* the points on temporal manifold $\mathcal{T}$ exist. As in the spatial case, one option is to reject the question as ill-formed. Metaphysically basic time points are part of the proposed *analysans* of temporal substantivalism but do not themselves form part of the *analysanda*. However, since temporal points are presumably not spatially located, either, treating them as non-temporal would put $\mathcal{T}$ firmly on the *abstract* side of the abstract/concrete divide, thus further eroding the difference between temporal substantivalism and modal accounts of time. The second option is to say that every time on $\mathcal{T}$ bears the exists-at relation E to itself, and to itself only. If existence at a time gets spelled out in terms of positions on $\mathcal{T}$ then every time should have the same temporal location as events happening then. But if that is right then each time only exists when it is present, and neither before nor after, and we get a succession of individual points rather than a complete temporal manifold.

Either option would allow us to avoid a temporal version of Zeno's regress, but neither of them would give us an account on which the temporal manifold forms a *substance* in the traditional sense. In the first case, time points are not temporal entities because they do not exist *at* any time. In the second case, they only exist for an instant. On neither view are time points continuants that persist through time, as required of a substance. By contrast, metaphysically basic spatial points do persist through time and $\mathcal{M}$ is made up of 'real stuff' in a fairly intuitive sense.

There is a second peculiarity worth noting. According to temporal substantivalism, the structural features of the manifold $\mathcal{T}$ determine how many events can occur, rather than the other way around. Instead of wondering, as we did in Section 2.1, whether there are enough events to generate a dense or complete time series, the question would be whether there are enough instants to permit the occurrence of so many events.

This makes a difference to Zeno's Dichotomy paradox, which 'asserts the non-existence of motion on the ground that that which is in locomotion must arrive at the half-way stage before it arrives at the goal' (*Physics* 239^b11). Suppose an object moves from point A to point B along some curve $\mathscr{C}$ on a continuous spatial manifold. In order for the object to get from A to B, it first has to get to the point H_0 that is located halfway between A and B on $\mathscr{C}$. But to get to H_0, it first has to get to the point H_{-1}, which is located halfway between A and H_0. And before it

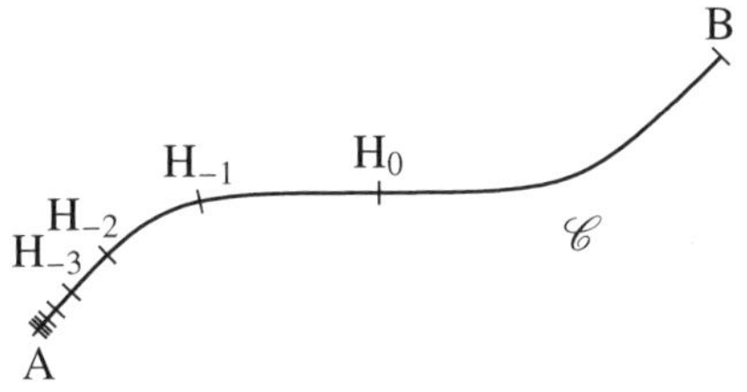

Figure 3.6. Zeno's Dichotomy

can get there, it first has to get to H_{-2}, with is located halfway between A and H_{-1}, and so on. There is an infinite sequence $\{H_{-i} : i = 0, 1, \ldots\}$ of points such that H_{-i} is halfway between A and H_{-i+1} and the object can only get to H_{-i} if it first gets to the halfway point H_{-i-1}. To get from A to B, the object would thus have to perform an infinite number of tasks. To get to the last position on the list $\ldots H_{-3}, H_{-2}, H_{-1}, H_0$, it has to get to all the preceding ones first, which seems as impossible as counting all the negative integers in order. There is no place to begin because there is no smallest negative integer, and too many numbers to count all of them. Zeno concludes that motion is impossible.

Most modern authors think that the Dichotomy is just one of the many puzzles about the infinite that were ultimately resolved by the rigorous theory of the continuum developed in the nineteenth century. But it would be hasty to discount the Dichotomy as a premature birth pang of the calculus, for it tells us something very interesting about the *number* of times. Motion is only possible if the cardinality of time at least matches that of space. To get from A at time t_1 to B at time t_2, there need to be at least as many instants between between t_1 and t_2 as there are points on the curve from A to B. So if space is a continuum then motion is only possible if there are also continuum many times.

For relationists about time, this is not a problem because the successive changes of the location of the moving object themselves would provide us with the required number of events to generate a continuum of instants. (Section 7.1 establishes a similar result for tense primitivism.) Substantivalists about time are in a rather different position. If the number of instants is indeed independent of what is happening within time, as they claim, then the possibility of motion would be a matter of metaphysical luck, or perhaps a sign of great benevolence on the part of the Creator, that He furnished our world with just enough times to allow us to move around. Unless temporal substantivalists could produce an argument that shows that there are necessarily as many times as places, they would be committed to the view that there are possible worlds whose inhabitants are condemned to immobility for lack of times.

Spatial and temporal substantivalists may also diverge on the question of whether the manifolds they postulate are composed of extensionless *points*. Some authors believe it to be part of our concept of space that any symmetric figure can be divided into two exactly symmetric halves. If that is right, then physical space cannot be isomorphic to the set $\mathbb{R}^3$ of all triples of real numbers. For example, it is impossible to divide a solid sphere in $\mathbb{R}^3$ into two exactly similar halves because only one of the two hemispheres can contain the dividing surface. To avoid this alleged difficulty, some authors propose a *pointless geometry*, according to which space has infinitely many sub-regions, but no minimal regions (i.e. points).[23] The same feature that some find troubling in the spatial case is desirable in the temporal case. Dividing the one-dimensional manifold $\mathcal{T}$ into two is to count some times as past and others as future. In this case, we do want there to be a dividing point to play the role of the present. This is the one respect in which it might be more natural to think of time as a continuous manifold than it is to think of space in this way, but this does little to lessen the weirdness of metaphysically basic time points.

[23] Such geometries are discussed in Carathéodory (1956), Grzegorczyk (1960), and Arntzenius (2008). Dummett (2000; 2005) rejects a continuum of extensionless time points for different reasons than the ones listed here; see also Meyer (2005a).

4

TENSE LOGIC

PRETENDING that times are like places is like trying to fit a square peg into a round hole. It can be done, but feels rather forced. Instead of pursuing spatial accounts of time any further, I now want to develop a modal account of time that treats instants on the model of the possible worlds familiar from modal logic. It turns out that the parallels between times and worlds are not perfect either, but I think this approach yields a far more attractive account of the metaphysics of time, without saddling us with any additional ontological commitments.

4.1 The property–time link

Event-relationism and temporal substantivalism both struggled to come up with a plausible solution to the Problem of Change, but I have not yet said anything about their relation to our folk theory of time. Spatial views of time derive at least some of their motivation from the pervasive use of spatial metaphors in ordinary temporal discourse. We speak of events in the *distant* past that we put *behind* us, or of those that used to be far in the future, but are now rapidly *approaching*. Many natural languages even use the same word to indicate the present moment and current location. This is true of 'present' in English, 'présent' in French, 'presente' in Italian, and also of 'gegenwärtig' in German. Spatial views of time want to take this kind of talk seriously.

Something similar is true for modal views of time; they just emphasize different aspects of our ordinary way of talking. Instead of focusing on spatial metaphors, they take their cue from the fact that, in most natural languages, verbs are tensed. In English, we cannot attribute a property to an object without committing ourselves, by our choice of tense, to a past, present, or future time at which the object possesses that attribute. This is what I want to call the *property–time link*. The questions of what an object is like and when it is thus are intimately connected in a way that is reminiscent of the modal dimension of property ascriptions. We also cannot say what an object is like without indicating, by our choice of modal auxiliary, whether we are talking about the actual or the merely possible possession of the property in question.

Advocates of spatial views of time tend to dismiss verb tense as a misleading idiosyncrasy of natural language:

> Our ordinary language shows a tiresome bias in its treatment of time. Relations of date are exalted grammatically as relations of position, weight, and color are not. This bias is of itself an inelegance, or breach of theoretical simplicity. Moreover, the form it takes—that of requiring that every verb form show a tense—is peculiarly productive of needless complications. (Quine 1960: 170)

Contrary to what Quine suggests, though, it is attempting to treat temporal distinctions like relations of position that leads to needless complications in our metaphysics and Section 3.1 was unable to come up with a plausible theory of time that does not end up vindicating ordinary language's 'tiresome bias' by endorsing the property–time link. This is a key difference between space and time. By saying that an object has a property we do not commit ourselves to a place where it has it. A material object might be located somewhere while it has a given property, but that is a conjunctive state of affairs. The object has the property *and* it is located at a particular place, but there is no property–location link that requires a spatial location for all property attributions.[24]

Even spatial views of time can be made to respect the property–time link, either by attributing properties to the temporal parts of objects, or by admitting only predicates with a temporal relatum. But none of these accounts seem particularly natural and they all incur extra ontological commitments. Modal views of time do things differently. They start out with the property–time link, rather than the mistaken belief that time is similar to space. Instead of pretending that time is a geometric space populated by material objects, they treat it as a kind of logical space at whose locations sentences or propositions take truth values.

Corresponding to their divergent views about temporal metaphysics, spatial and modal views of time offer different accounts of temporal logic. While spatial views try to account for temporal distinctions within a standard first-order logic, modal views of time advocate an extension of this framework with intensional tense operators. One popular proposal uses a past tense operator P ('it was the case that'), a future tense operator F ('it will be the case that'), and reserves sentences without such operators for making claims about the present time.

[24] This is what is behind the failure of the modal theories of space discussed on p. 26.

Modal views of time therefore have two main components: *tense operators* like P and F and a collection of *times* that serve as indices at which sentences take truth value. This leads to the question of which of the two enjoys priority over the other. According to the *tense primitivism* that I advocate, all temporal notions are to be analysed in terms of conceptually primitive tense operators. This includes instants of time, which Chapter 6 identifies with maximal consistent sets of sentences of tense logic. By taking P and F as conceptually primitive, I do not mean to advance a psychological thesis about concept acquisition. Perhaps there are some concepts, such as shapes and colours, that are definitionally basic in that they can only be defined ostensively, rather than in terms of a stipulative definition whose definiens already takes some other concepts for granted. That is not what is at issue here. To say that they are conceptually primitive is to make a claim about the theoretical role of tense operators. My modal theory of time provides information about how tense operators interact with other logical constants, but it does not try to explain them in terms of anything else.

I am well aware that it is a bit unorthodox to take intensional operators, rather than predicates and relations, as conceptual primitives, but I do think that time is a special case that requires special treatment. The task of the next few chapters is to lay out the main components of such a modal theory of time, and to dispel some of the more common prejudices against intensional theories in fundamental metaphysics.

4.2 Modal logic in a nutshell

To explain how tense operators work, let me begin with a quick survey of the modal logic that serves as the inspiration of my theory of time. For this purpose, it is helpful to construct a system of modal logic from the bottom up. At the base level, we have the logic of sentence letters. Its formal language only contains atomic sentence letters A, B, C, and D, with and without numerals as subscripts.[25] A model for this language is a map $\mathfrak{m}$ that assigns truth values T ('true') and F ('false') to all sentences. An atomic sentence φ is said to be true in a model, written $\mathfrak{m} \models \varphi$, if and only if $\mathfrak{m}(\varphi) = \text{T}$. Since the truth value assigned to one atomic sentence imposes no restrictions on the values a model assigns to other sentences, there are no interesting logical relations between the sentences of the language of sentence letters.

[25] Use and mention are freely confused throughout the text, except for the rare occasions when it is unclear from the context whether I am talking about a name or its referent.

To get a non-trivial logic, we need to enlarge our language with logical constants. In a first step, we add the sentence connectives $\neg$ ('not') and $\rightarrow$ ('if ... then ...') and stipulate that both $\neg\varphi$ and $\varphi \rightarrow \psi$ are (molecular) sentences if φ and ψ are sentences. Models for this extended language are the same as before, assignments of truth-values to atomic sentences, but we now extend the true-at relation $\models$ to sentences that contain the two new logical constants:

$$\begin{array}{lll} \mathfrak{m} \models \varphi & \text{iff} & \varphi \text{ is atomic and } \mathfrak{m}(\varphi) = \mathrm{T} \\ \mathfrak{m} \models \neg\varphi & \text{iff} & \text{it is not the case that } \mathfrak{m} \models \varphi \\ \mathfrak{m} \models \varphi \rightarrow \psi & \text{iff} & \text{either } \mathfrak{m} \models \neg\varphi \text{ or } \mathfrak{m} \models \psi \end{array}$$

A sentence φ is said to be a logical consequence of premises $\psi_1, \ldots, \psi_n$ just in case φ is true in all models in which all of the premises are true. The sentence φ is a logical truth if it is true in all models, and φ and ψ are logically equivalent if they have the same truth values in all models.

These clauses ensure that $\neg$ and $\rightarrow$ are truth-functional connectives that always take the same truth value on sentences with the same truth values. The truth value of a molecular sentence is always a function of the values assigned to its atomic components, and we can summarize the contributions made by these two connectives in terms of a truth table:

φ	ψ	$\varphi \rightarrow \psi$	$\neg\varphi$
T	T	T	
T	F	F	F
F	T	T	T
F	F	T	

There are many other truth-functional sentence connectives that one could consider at this point, but all of them can be defined in terms $\neg$ and $\rightarrow$. For example, we can treat $\varphi \vee \psi$ ('Either φ or ψ') as an abbreviation for $\neg\varphi \rightarrow \psi$. The standard connectives $\wedge$ ('and') and $\leftrightarrow$ ('if and only if') can be defined in a similar way.

The model theory characterizes $\neg$ and $\rightarrow$ semantically, in terms of the truth values that sentences containing them take in the different models of our language. We can also define their logical properties syntactically, by using a proof theory with axioms and inference rules:

A1 $\varphi \rightarrow (\psi \rightarrow \varphi)$
A2 $(\neg\varphi \rightarrow \neg\psi) \rightarrow (\psi \rightarrow \varphi)$
A3 $(\varphi \rightarrow (\psi \rightarrow \omega)) \rightarrow ((\varphi \rightarrow \psi) \rightarrow (\varphi \rightarrow \omega))$

MP From $\varphi \rightarrow \psi$ and φ infer ψ

A proof of φ from premises $\psi_1, \ldots, \psi_n$ is an ordered sequence of sentences that ends with φ such that each element of the sequence is either one of the premises, an instance of one of the axiom schemata A1–A3, or follows from preceding sentences by the rule of *modus ponens* MP. If there is such a proof then we write $\psi_1, \ldots, \psi_n \vdash \varphi$ and say that φ is provable from these premises. A sentence is a theorem of our system, $\vdash \varphi$, if it is provable from zero premises and thus follows by repeated application of MP from the three logical axioms alone.

According to the soundness and completeness theorem for propositional logic, a sentence is provable from a set of premises in our axiomatic system if and only if it is one of their logical consequence. The semantic and the syntactic characterization of the logical properties of $\neg$ and $\rightarrow$ thus yield the same result.

We can think of the models for propositional logic as different ways the world might have been; that is, as different possible worlds. As it stands, our logic only permits us to make claims about one of these 'worlds' at a time, which is being presented as the actual world. To make claims about how things are in other possible worlds (i.e. in other models), we introduce the sentential operator $\Diamond$ ('possibly') as an additional logical constant. It is also convenient to treat $\Box\varphi$ ('necessarily, φ') as an abbreviation for $\neg\Diamond\neg\varphi$. A model for such a modal logic then consists of nothing but a set $\mathfrak{W}$ of models of the underlying propositional logic. The only difference is that we now call these models 'possible worlds' and pick one of them as 'actual world' $\mathfrak{i}$. We also need a binary *accessibility relation* $\mathfrak{R}$ on the set of propositional models $\mathfrak{W}$ that specifies which worlds are possible relative to a given world.

Given such a model $\langle \mathfrak{W}, \mathfrak{R}, \mathfrak{i} \rangle$ of modal logic, the true-at relation $\models$ between worlds and sentences is defined just as for propositional logic, except for an additional clause to govern the new modal operator:

$$
\begin{array}{lll}
\mathfrak{w} \models \varphi & \text{iff} & \varphi \text{ is atomic and } \mathfrak{w}(\varphi) = \mathrm{T} \\
\mathfrak{w} \models \neg\varphi & \text{iff} & \text{it is not the case that } \mathfrak{w} \models \varphi \\
\mathfrak{w} \models \varphi \rightarrow \psi & \text{iff} & \text{either } \mathfrak{w} \models \neg\varphi \text{ or } \mathfrak{w} \models \psi \\
\mathfrak{w} \models \Diamond\varphi & \text{iff} & \mathfrak{w}' \models \varphi \text{ for some } \mathfrak{w}' \in \mathfrak{W} \text{ such that } \mathfrak{R}\mathfrak{w}\mathfrak{w}'
\end{array}
$$

A sentence φ is true in a model of modal logic just in case it is true at its actual world, $\mathfrak{i} \models \varphi$. In particular, this means that $\Diamond\varphi$ is true in a model if φ is true in some possible world that is accessible from the actual world of the model, and $\Box\varphi$ is true if φ is true in all such worlds.

As for propositional logic, the logical properties of $\Diamond$ can also be characterized axiomatically, but in this case there is a whole range of different systems one could consider. One popular choice is the modal system S5, which can be obtained by adding the following axioms and inference rule to the propositional logic composed of A1–A3 and MP:

M1	$\Box(\varphi \rightarrow \psi) \rightarrow (\Box\varphi \rightarrow \Box\psi)$
M2	$\Box\varphi \rightarrow \varphi$
M3	$\Box\varphi \rightarrow \Box\Box\varphi$
M4	$\Diamond\varphi \rightarrow \Box\Diamond\varphi$
NEC	If $\vdash \varphi$ then $\vdash \Box\varphi$

According to these new axioms, ψ is necessary if both $\varphi \rightarrow \psi$ and φ are necessary (M1); necessary truths are actually true (M2); necessary truths are *necessarily* necessary (M3); and φ is necessarily possible if it is possible (M4). The *necessitation rule* NEC asserts that all theorems of our modal logic are themselves necessary.

The modal operator $\Diamond$ has similar syntactic properties as the negation symbol $\neg$. Both are unary sentence connectives that yield another sentence when prefixed to a sentence. The main difference between them concerns their semantic properties. Suppose that φ and ψ are both false in a given model of modal logic. Then $\neg\varphi$ and $\neg\psi$ are both true in that model, but the truth values of $\Diamond\varphi$ and $\Diamond\psi$ can still differ. For example, suppose that φ is false in all other worlds of the model as well, but that ψ is true in some of them. In that case, $\Diamond\varphi$ would be false and $\Diamond\psi$ true even though the embedded sentences have the same actual truth value. Unlike negation, modal operators are thus not truth-functional; they are *intensional* connectives that are only guaranteed to deliver the same truth values on logically equivalent sentences.[26]

Propositional logic has one soundness and completeness theorem that shows that our model theory and our proof system are appropriately matched. The modal case has a more nuanced structure. There is a sequence of such theorems that demonstrate how adding the axioms M1–M4 to our propositional logic imposes successively more stringent restrictions on the accessibility relation $\mathfrak{R}$ between possible worlds.[27]

Intensionality. The basic modal system K consists of axioms A1–A3, rules MP and NEC, and axiom M1. This system requires $\Diamond$ to

[26] See also Williamson (2006); Cresswell (1970) discusses *non*-intensional operators.

[27] For proofs, see Hughes and Cresswell (1996).

be an intensional connective, by ensuring that $\Diamond\varphi$ and $\Diamond\psi$ are provably equivalent whenever φ and ψ are. Its theorems are those sentences that are true in all models, without restrictions on $\mathfrak{R}$.

Reflexivity. By adding M2 to K, we get the system T, which ensures that the necessity operator $\Box$ is factive and only applies to sentences that are true in the actual world. The theorems of T are those sentences that are true in all models in which the accessibility relation $\mathfrak{R}$ is reflexive and in which $\mathfrak{R}\mathfrak{w}\mathfrak{w}$ for all $\mathfrak{w} \in \mathfrak{W}$.

Transitivity. The system S4 is obtained by adding M3, which requires that all true claims of the form $\Box\varphi$ are themselves necessary. The theorems of S4 are those sentences that are true in all models in which $\mathfrak{R}$ is reflexive and transitive; that is, for which $\mathfrak{R}\mathfrak{w}\mathfrak{w}''$ holds whenever $\mathfrak{R}\mathfrak{w}\mathfrak{w}'$ and $\mathfrak{R}\mathfrak{w}'\mathfrak{w}''$ are true.

Symmetry. The complete system S5 contains M4 in addition to everything else. Axiom M4 is a complement to M3 and requires that also all true claims of type $\Diamond\varphi$ are necessary. To ensure this, $\mathfrak{R}$ must also be symmetric, and $\mathfrak{R}\mathfrak{w}\mathfrak{w}'$ hold whenever $\mathfrak{R}\mathfrak{w}'\mathfrak{w}$.

Since we are primarily concerned with the nature of time, we need not settle which of these systems, if any, is the correct theory of possibility. I am inclined to accept S5, which is sound and complete with respect to all models in which $\mathfrak{R}$ is an equivalence relation; that is, is reflexive, transitive, and symmetric. In this case, every possible world is accessible from every other one. We can drop the redundant accessibility relation altogether and say that $\Diamond\varphi$ is true in a model of S5 just in case φ is true in some possible world of that model.

4.3 Tense operators

Modal theories of time propose to treat temporal distinctions in a similar way, by introducing suitable tense operators.[28] The easiest way of doing this is by reinterpreting the modal system S5 as a tense logic. Suppose we follow Arthur Prior (1957a) and read $\Diamond$ as 'sometimes', $\Box$ as 'always', and reserve sentences without such operators to make claims about what is presently the case. Then axiom M1 claims that ψ is always true if $\varphi \rightarrow \psi$ and φ are always true; M2 that what is always true is true now; M3 that what is always true is *always* always true;

[28] Introductions to tense logic are in van Benthem (1983), Burgess (2002; 2009), and McArthur (1976). Pnueli (1977) discusses applications of tense logic in computer science.

and M4 that what is sometimes true is always sometimes true. NEC becomes a *temporal generalization* rule that claims that logical truths are always true. The model theory can be recycled in a similar manner, by calling the elements of $\mathfrak{W}$ 'times', rather than 'possible worlds', and $\mathfrak{i}$ 'the present' rather than 'the actual world'. We have one formalism that can serve either as a system of modal logic or as a simple tense logic:

Formalism	Modal logic	Tense logic
$\Diamond$	possibly	sometimes
$\Box$	necessarily	always
$\mathfrak{W}$	possible worlds	times
$\models$	true in a world	true at a time
$\mathfrak{i}$	actual world	the present

This illustrates some of the formal similarities between tense and modal logic, but it also ignores the distinction between past and future, which has no modal analogue. To account for this additional structure, systems of tense logic typically use two different 'possibility' operators. The standard choice, which I will adopt here, is to employ a past tense operator P ('it was the case that') and a future tense operator F ('it will be the case that'). Their dual operators H ('it has always been the case that') and G ('it will always be the case that') are defined as abbreviations for $\neg\mathsf{P}\neg$ and $\neg\mathsf{F}\neg$, respectively. A very simple P/F-tense logic can then be obtained by adding the following axiom schema and inference rule to our basic propositional logic:

Z1	$\mathsf{H}(\varphi \to \psi) \to (\mathsf{H}\varphi \to \mathsf{H}\psi)$	$\mathsf{G}(\varphi \to \psi) \to (\mathsf{G}\varphi \to \mathsf{G}\psi)$
TG	If $\vdash \varphi$ then $\vdash \mathsf{H}\varphi$	If $\vdash \varphi$ then $\vdash \mathsf{G}\varphi$

The resulting tense logic consists of two copies of the simple modal system K; there is one copy for the past and one for the future. The tense operators P and F play the role of 'possibility' operators and H and G that of 'necessity' operators. The axiom schema Z1 is a tense version of M1 and the *temporal generalization* rule TG performs the function of NEC, reflecting the fact that theorems of tense logic are always true.

This system is strong enough to guarantee that P and F are intensional connectives, but it is a little bit too weak to serve as a plausible tense logic. We do not want the models for tense logic to look quite like doubled-up models for modal logic. To do so would require two independent temporal order relations: an earlier-than relation $<$ to serve as an ‘accessibility relation’ for P and a later-than relation > to perform the same function for F. One of the very few uncontroversial claims about temporal order is that < and > are not independent, but are converses of one another. If one time is earlier than a second, then the latter is later than the former. To enforce this, we need an additional axiom schema:

$$\text{Z2} \quad \varphi \rightarrow \mathsf{HF}\varphi \quad \varphi \rightarrow \mathsf{GP}\varphi$$

Let Z (‘*Zeitlogik*’) be the resulting system of tense logic. As of now, this system consists of a standard propositional logic A1–A3 and MP plus the tense axioms Z1 and Z2 and the temporal generalisation rule TG. (Chapter 8 adds rules and axioms for quantifiers and identity.)

A model for Z is a triple $\langle \mathfrak{T}, <, \mathfrak{p} \rangle$, where $\mathfrak{T}$ is a set of models of the underlying propositional logic that we call ‘times’ whenever they occur in this role. One of these times is chosen as ‘the present’ $\mathfrak{p}$. There is also an earlier-than relation < on $\mathfrak{T}$ that functions as accessibility relation for the tense operators. Similar to the model theory for modal logic, the true-at relation $\models$ between times and sentences is defined via:

$\mathsf{t} \models \varphi$	iff φ is atomic and $t(\varphi) = \mathrm{T}$
$\mathsf{t} \models \neg\varphi$	iff it is not the case that $\mathsf{t} \models \varphi$
$\mathsf{t} \models \varphi \rightarrow \psi$	iff either $\mathsf{t} \models \neg\varphi$ or $\mathsf{t} \models \psi$
$\mathsf{t} \models \mathsf{P}\varphi$	iff $\mathsf{t}' \models \varphi$ for some $\mathsf{t}' \in \mathfrak{T}$ such that $\mathsf{t}' < \mathsf{t}$
$\mathsf{t} \models \mathsf{F}\varphi$	iff $\mathsf{t}' \models \varphi$ for some $\mathsf{t}' \in \mathfrak{T}$ such that $\mathsf{t} < \mathsf{t}'$

A sentence φ is true in a model of tense logic if and only if it is true at its present time, $\mathfrak{p} \models \varphi$. Logical consequence, logical truth, and logical equivalence are defined as before, in terms of truth in a model. A small modification of the proof of the soundness and completeness theorem for K to take account of Z2 then allows us to show that a sentence is a theorem of the tense logic Z if and only if it is true in all models, irrespective of the properties of the earlier-than relation.

To impose more specific constraints on <, there is a range of further axioms one could consider. Similar to the modal case, we get an entire sequence of different soundness and completeness theorems:[29]

Transitivity. To ensure that the earlier-than relation is transitive, we could add $\mathsf{PP}\varphi \rightarrow \mathsf{P}\varphi$ and $\mathsf{FF}\varphi \rightarrow \mathsf{F}\varphi$. The theorems of this

[29] See Cocchiarella (1966), Prior (1966), Gabbay (1975), and Burgess (2002). One property that is absent from this list is the *antisymmetry* of the earlier-than relation, which requires $\mathsf{t} < \mathsf{t}'$ to be false whenever $\mathsf{t}' < \mathsf{t}$ is true, and which cannot be captured by an axiom schema. Section 7.2 discusses this issue in more detail.

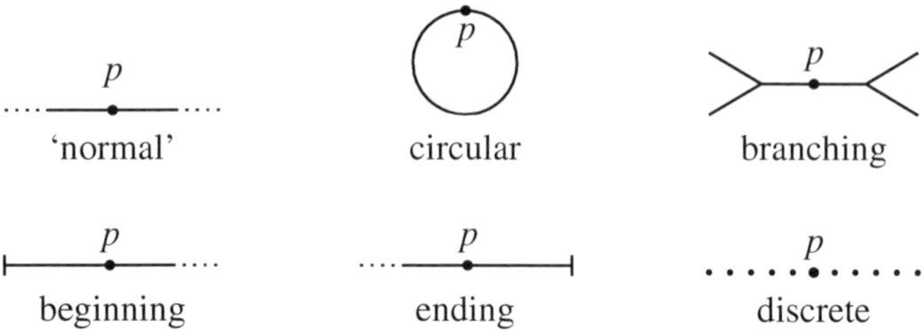

Figure 4.1. Different time series

extended system are those sentences that are true in all models for which $t_1 < t_3$ whenever both $t_1 < t_2$ and $t_2 < t_3$.

Comparability. The system Z permits the time series to *branch*. In a forward branch, all times earlier than the time of branching are earlier than all times in the branches, but no time in one branch is earlier than any time in another. We can prohibit forward branches with $(\mathsf{F}\varphi \wedge \mathsf{F}\psi) \rightarrow (\mathsf{F}(\varphi \wedge \psi) \vee \mathsf{F}(\varphi \wedge \mathsf{F}\psi) \vee \mathsf{F}(\mathsf{F}\varphi \wedge \psi))$; backward branches can be ruled out by adopting $(\mathsf{P}\varphi \wedge \mathsf{P}\psi) \rightarrow (\mathsf{P}(\varphi \wedge \psi) \vee \mathsf{P}(\varphi \wedge \mathsf{P}\psi) \vee \mathsf{P}(\mathsf{P}\varphi \wedge \psi))$. If we adopt both axioms then $<$ must be comparable in the sense that one of two different instants of time is always earlier than the other.

Endpoints. To rule out the existence of a first or last time, we use $\varphi \rightarrow \mathsf{PF}\varphi$ or $\varphi \rightarrow \mathsf{FP}\varphi$. We can guarantee the existence of a first time with $\mathsf{H}\bot \vee \mathsf{PH}\bot$ and the existence of a last time with $\mathsf{G}\bot \vee \mathsf{FG}\bot$. Here $\bot$ stands for an arbitrary but fixed logical falsehood.

Denseness. A dense time series, in which there is always another time between any two given ones, can be enforced with $\mathsf{P}\varphi \rightarrow \mathsf{PP}\varphi$ and $\mathsf{F}\varphi \rightarrow \mathsf{FF}\varphi$. The axioms $(\varphi \wedge \mathsf{H}\varphi) \rightarrow \mathsf{FH}\varphi$ and $(\varphi \wedge \mathsf{G}\varphi) \rightarrow \mathsf{PG}\varphi$ yield a *discrete* time (without endpoints) in which all times have immediate predecessors and successors.

Completeness. To guarantee a complete time series, we add the axioms $(\mathsf{F}\varphi \wedge \mathsf{FG}\neg\varphi) \rightarrow \mathsf{F}(\mathsf{HF}\varphi \wedge \mathsf{G}\neg\varphi)$ and $(\mathsf{P}\varphi \wedge \mathsf{PH}\neg\varphi) \rightarrow \mathsf{P}(\mathsf{GP}\varphi \wedge \mathsf{H}\neg\varphi)$. This ensures that every non-empty subset of $\mathfrak{T}$ with an upper (lower) bound has a least upper (greatest lower) bound.

Different systems of tense logic combine these axioms in different ways, so there is a wide range of modal theories of time to choose from. These theories differ with respect to which features of $<$ they take to be part of the nature of time itself, and which features they regard as part of what is happening within time. My proposal is that we take the very weak tense logic Z as our fundamental theory of the nature of time.

If we grant that the theorems of Z are analytic, this would also mean that our knowledge of time is analytic a priori. This is a far less radical thesis than it might seem at first. Mine is a *minimal* theory of time, according to which there is very little to be known about time itself. Most of what rival accounts take to be part of nature of time, my view regards as a feature of what is happening within time. The simple tense logic Z does not prejudge whether there is such a thing as absolute simultaneity, how many times there are, whether time has a metric structure, or whether it branches. I think this is an attractive feature because these are empirical questions that should have no place in an investigation of the nature of time itself. It also means that my view and rival accounts largely agree *which* questions are empirical. Where we disagree is on whether these are questions about the nature of time or questions about what is happening within time.

When I introduced tense and modal operators, I suggested that we think of them as new logical constants. Nothing of importance depends on whether they really qualify for this role. On one popular view, an expression counts as a logical constant if its meaning is exhausted by its inferential role. This works reasonably well for propositional logic, where one can prove that there are unique truth-functional connectives that play the inferential role assigned to $\rightarrow$ and $\neg$ by the axioms A1–A3 and the inference rule MP. This result does not extend to tense and modal operators. For instance, the system S5 does not specify a unique intensional operator because it does not tell us whether to read $\Diamond$ as the modal operator 'possibly' or as the tense operator 'sometimes'.

It is not clear whether this shows that tense and modal operators are not logical constants, or whether it merely indicates that the proposed account of logical constanthood is too narrow since it fails to admit them. I shall not worry about this question here because I do not think there is an interesting philosophical question about the boundaries of logic, anyway. As Ian Hacking notes, 'there is no point in trying to separate logic from other science, except for that profound speculation about the nature and origin of necessary truth' (1994: 3). What we make of this observation, though, depends on what we think about W. V. Quine's (1951) claim that the distinction between analytic and synthetic truths is untenable. Quine's view is that all truths are of the same kind. Very roughly, he thinks that they are all synthetic. If he is right about this then it would be futile to look for a principled distinction between logical and other truths. If Quine is wrong, though, then it is clear on which side of the divide logical truths fall: they are analytic if anything is. An account of logicality would attempt to make a further distinction amongst the analytic truths. But if logical truths are analytic then the source of their necessity has already been identified: they are true in virtue of meaning. Hence the boundaries of logic do not raise any interesting philosophical problems in addition to the analytic–synthetic distinction. I therefore plan to be generous with the label 'logic' and award it to any systematic study of truths, including my tense logic Z.

5

PROPOSITIONS

If we take the tense operators P and F as conceptually primitive then our theory of time describes reality in terms of tensed sentences that take different truth values at different time. This raises the question of what implications such a theory has for propositions, which are expressed by sentences. It is widely thought that modal theories of time are committed to the existence of *temporal* propositions, which change their truth values over time, while spatial theories require *eternal* propositions, whose truth values are always the same. I think this view is mistaken. Modal theories of time might provide a hospitable environment for temporal propositions but they do not require them.

5.1 Contents of assertion

What sort of propositions we must accept depends on what we want to do with them. There are three main areas where propositions find employment: as the contents of assertions, as the objects of belief and other attitudes, and in an account of facts. In all three applications, propositions appear as the bearers of truth values. Let me therefore start with the paradigm example of something that can be said to be true or false: what is being asserted by an utterance of an indicative sentence.

For sake of argument, let me grant that what is being asserted by such utterances are propositions and let me simplify matters by ignoring all aspects of the context of utterances other than the time of their occurrences. Then the question is whether what is being asserted can change its truth value over time (temporal proposition), or whether its truth value is always constant (eternal proposition). We can get a rough answer to this question by considering what sentences one would have to utter at two different times to assert the same proposition on both occasions. In the language of tense logic, the same sentence can be used to assert a true proposition at one time and a false one at another. Suppose that Adam is kvetching at t_1 but not at t_2. Then an utterance of the sentence Ka expresses a true proposition if it occurs at t_1 and it expresses a false proposition if it occurs at t_2. Since what is being asserted differs in truth value, the two utterances cannot express the same eternal proposition, but they could express the same temporal proposition, which changes its truth value from t_1 to t_2.

Say that a sentence *characterizes* a proposition if and only if every utterance of the sentence would express that proposition, irrespective of when it occurs. Since we are ignoring all features of the context other than time, every utterance of a sentence expresses some proposition, but a sentence only characterizes a proposition if its temporal structure appropriately matches the propositions that happen to exist. If there were temporal propositions then the tensed sentences of tense logic would characterize some of them, but many of them would fail to characterize eternal propositions. The matter is the reverse for spatial theories, which describe the time series in a tenseless manner. Temporal substantivalists would use Kat_1 to describe Adam's activity at time t_1. If this sentence can be uttered truly at one time then it can always be uttered truly, including the time t_2 at which Adam is no longer kvetching. Whereas tensed sentences would characterize temporal propositions (if there were such propositions) tenseless sentences would characterize eternal propositions (if there were such propositions).[30]

This is a noteworthy difference between spatial and modal theories, but it is not clear that it tells us anything interesting about the metaphysics of time. Without further argument, there is no reason to suppose that our theory of time must also serve as our theory of the content of assertions, or vice versa. One might well be satisfied with a theory in which the same proposition gets picked out by different sentences at different times. Indeed, such a theory might even be far more informative about the temporal features of propositions than one presented in terms of sentences that characterize propositions.

Moreover, if we regard propositions as abstract objects, as many philosophers of language do, then there seems to be little point in agonizing about whether to accept temporal or eternal propositions. Since abstract objects do not take up any space, we could just accept *both* eternal and temporal propositions and be done with it.

5.2 Beliefs and other attitudes

Similar remarks apply to the employment of propositions in an account of the objects of belief and other attitudes. Suppose that, at some time

[30]This mirrors the distinction between chronologically 'definite' and chronologically 'indefinite' statements in Rescher (1966).

t_1, Claire believes that Adam is kvetching then and that she does not change her mind until some later time t_2. That is, Claire believes the same things at t_2 as she did at t_1. The question is how to describe her beliefs at the two times. If Claire believes an eternal proposition, as seems plausible in this case,[31] then temporal substantivalists could say that, at time t_1, Claire believes that Kat_1. They could use the same sentence to express Claire's belief at time t_2, when she still believes that Kat_1. However, the object of her belief can also be described in tensed terms, by using different sentences at the two times. At time t_1, tense primitivists could say that Claire believes that Ka. They cannot use the same tensed sentence to express the object of her unchanged beliefs at t_2, but they could describe it by saying that she believes that $\mathsf{P}Ka$.

Neither account is entirely satisfactory. Temporal substantivalists risk ascribing too much belief to Claire. Suppose that, at time t_1, Claire does indeed believe that Adam is kvetching, but that she has no idea what time it is. By saying that she believes that Adam is kvetching at t_1 we would be turning her belief *at* a time into a belief *about* that time and thus ascribe a belief to Claire that she does not in fact possess.[32] Claire's belief might be better described by including a suitable auxiliary claim that she believes to have been true while Adam was kvetching. For instance, if she believes that Adam was kvetching while the banana was yellow, then the weaker proposition that $\exists t\,(Kat \wedge Ybt)$ might give us a better account of her belief than Kat_1.

This does not mean that Kat_1 fails to characterize the proposition that would have been expressed if Claire had actually uttered 'Adam is kvetching' at t_1. Our account of the content of assertions might well come apart from our account of the objects of beliefs. Just as an agent need not assert what she believes, she need not believe what she asserts. Claire might well believe that $\exists t\,(Kat \wedge Ybt)$, but this is clearly not what would be asserted by her utterance of 'Adam is kvetching' at t_1. What proposition an utterance expresses can depend on features of the context that are unknown to the utterer.

The account offered by tense primitivists suffers from the opposite

[31] Plausible, but not uncontroversial. See Hamblin (1970), Richard (1981), Salmon (1986; 1989), Aronszajn (1996), Fitch (1998), Cappelen and Hawthorne (2009), and Brogaard (2012).

[32] This is a version of what Perry (2001) calls the temporal knowledge argument. See also Prior's (1962) argument that an eternal God, who exists 'outside' time altogether, could not be omniscient because he would not know what time it is.

problem. It runs the risk of attributing too little belief. At time t_2, Claire might indeed believe that PKa, but it is likely that she believes something far more specific than that. Also in this case, the introduction of a suitable auxiliary claim α could improve matters. For instance, α might describe the hands on Claire's watch, or it might report some event that happened while Adam is believed to have been kvetching. This would allow tense primitivists to characterize Claire's belief at t_2 more finely, by saying that she believes that P$(Ka \wedge \alpha)$.

There are also cases in which the objects of attitudes seem more naturally characterized in terms of temporal propositions. Suppose that, at time t_1, I *dread* that my dentist will pull out all my teeth. I no longer dread this at the later time t_2 because the dentist did as I feared and I have no teeth left to be pulled out. I am now *relieved* that the operation is over. In this case, it seems wrong to say that what I dread at t_1, and what I am relieved about at t_2, is that the dentist pulled out my teeth at some intermediate time. If the object of dread is a proposition, it would be more plausible to say that I have changing attitudes towards a temporal proposition that is characterized by the sentence 'The dentist will pull out my teeth'. At t_1, I dreaded this because the proposition was true then, and I no longer dread it at t_2 because the proposition has become false. If that is the right way of characterizing such cases then at least some objects of attitudes are temporal propositions.[33]

How the objects of belief and dread are related to the contents of assertions is an interesting issue, but it is a question that a theory of propositional attitudes needs to answer, not a theory of time. Since modal and spatial accounts of time can easily accept either type of proposition, there is no reason why they ought to concern themselves with these problems. The contrast between temporal and eternal propositions is one that involves time, but is not about the nature of time itself.

5.3 Facts and truth-making

The employment of propositions in an account of facts raises slightly different issues. Very roughly, the idea is that 'the fact that φ' and 'the fact that ψ' pick out the same fact just in case the sentences φ and ψ pick

[33]There is an extensive literature on this issue, and especially on the question of whether the objects of relief are tensed propositions. See, e.g. Prior (1959), Mellor (1981b; 1983), MacBeath (1983), Garrett (1988), Richard (2003), and Higginbotham (2003). Merricks (2009) denies that fears and desires are propositional attitudes.

out the same true proposition.[34] The distinction between temporal and eternal propositions would thus project onto a corresponding distinction amongst facts. A *tensed* fact is a fact that corresponds to a true temporal proposition, and a *tenseless* fact one that corresponds to an eternal one. Since temporal propositions change truth values over time, and since only true propositions pick out facts, tensed facts exist at some times but not at others, while tenseless facts are always the same.

Contingent propositions are not true on their own, but are so in virtue of the way the world is; *truth supervenes on being*, we are told. One popular way of spelling out this idea is in terms of the *truthmaker principle*. This principle requires every true proposition to possess a truthmaker, which is an object whose *existence* alone guarantees the truth of that proposition.[35] As it happens, many advocates of the truth-maker principle also believe that facts are the only plausible candidate for truthmakers. If that is right, then the central metaphysical question about time is whether the facts are tensed or tenseless, and not whether to adopt a spatial or a modal theory.

If we think about the philosophy of time in these terms, though, then we are unlikely to get decisive arguments on either side. What sort of facts we get depends on what sort of propositions we postulate, yet if the main employment for propositions is in an account of facts, rather than in accounts of assertion or attitudes, then we cannot get an argument for or against tensed facts that is not begging the question. Kit Fine (2005c: sec. 1) concludes that the debate only makes sense against the backdrop of what he calls a 'distinctively metaphysical concept of reality' that is concerned with how things *really* are. I do not see how this avoids the circularity problem, but that is not an issue I want to press. I propose that we *reject* the truthmaker principle, in which case such problems do not arise in the first place.

Proponents of the truthmaker principle often tout its ability to catch *cheaters*, who are philosophers that claim to give an account of a certain subject matter without wanting to pay the ontological price. Arm-

[34]This simple picture needs to be complicated a little bit if we want to avoid commitment to disjunctive or negative facts. See Bigelow (1988: 130–3) and Lewis (1992: 215–19). We also need to be careful in specifying the conditions under which two sentences pick out the same proposition, lest we become vulnerable to Davidson's (1980c: 117–18) infamous argument that there is only one fact. See Bennett (1988: secs. 15, 16).

[35]A prominent advocate of this view is Armstrong (1997: ch. 8). See also Mellor (1998: ch. 2), Dodd (2002), and Merricks (2007). A survey of the recent debate about truthmakers can be found in Beebee and Dodd (2005) and Rodriguez-Pereyra (2006).

strong's favourite example of cheating is the dispositional theory of the mind offered by Gilbert Ryle (1949). On Ryle's view, being in a certain mental state consists of nothing more than being disposed to act in a certain way when put in certain circumstances. While Armstrong agrees that dispositions play a key role in the philosophy of mind, he does not think that this is the end of the matter: 'we need then to go on to consider the question of the truthmaker for these dispositional truths. What is there in the world in virtue of which these truths are true? Ryle had no answer' (2004: 3). It is not obvious to me, though, that Ryle must answer this question. Armstrong makes it clear in (1996) and elsewhere that *he* does not think that there could be a dispositional property that does not possess a categorical basis, which is a cluster of non-dispositional properties that accounts for the presence of the disposition. His real complaint seems to be that Ryle postulates dispositions without bothering to identify such bases. Yet the claim that all dispositions must have a categorical basis is a decidedly metaphysical thesis, and one that many contemporary philosophers reject.[36] Ryle's cheating appears to amount to little more than a failure to anticipate Armstrong's own theory of dispositions.

Something like this holds more generally. Instead of ensuring metaphysical honesty, the truthmaker principle tends to generate more extravagant philosophical views than it eliminates.[37] Consider how the principle forces us to treat simple property attributions such as 'The apple is red'. Since the apple could have failed to be red, its existence alone does not guarantee the truth of the sentence. So the apple is not a truthmaker. Since redness could have existed without being instantiated by the apple, redness is not a truthmaker, either. And since the apple and redness can both exist without the former instantiating the latter, the mereological sum of the apple and redness, or the set containing both of them, also do not qualify as truthmakers. Even in such elementary cases, we only get truthmakers if we accept some rather fancy metaphysics. Armstrong favours an ontology of facts, which are a way of marrying properties to objects. The *fact that the apple is red* cannot exist without both the apple and redness, and without the former instan-

[36] See Tooley (1972), Mellor (1974), Meyer (2002), and McKitrick (2003).

[37] Apart from problems that arise when applying the truthmaker principle, there are also questions about how to formulate it in the first place. It is not obvious how to account for necessary truths or negative existential claims, and Williamson (1999) argues that the principle conflicts with a plausible quantified modal logic.

tiating the latter. If such a fact exists then it qualifies as a truthmaker for 'The apple is red'. Other possibilities are to follow Kevin Mulligan et al. (1984) and adopt an ontology of tropes, or to espouse the radical thesis that no object has any properties contingently, in which case the mere existence of the apple would already yield its redness.

David Lewis (1992: 216) makes a useful distinction between two views about truth-making. One is that truth depends on *how* objects are, the other that it depends on *whether* they are. The former view might be an uncontroversial truism, but the truthmaker principle demands a truth-maker in the latter sense and the right thing to say about 'The apple is red' is surely that its truth depends on *how matters are with the apple*. Only somebody who is already committed to facts or tropes would think otherwise. The truthmaker principle codifies a preference for a certain type of metaphysical theory rather than providing a neutral court of appeal in which rival philosophical views can be adjudicated.

Another reason for rejecting the truthmaker principle arises from my theory of time itself. Quentin Smith complains that Prior 'never adequately clarified the positive ontological import of his tense operators' (1994: 11). But to demand an extensional theory with transparent ontological commitments that spells out the 'deep truth' behind an intensional theory of time is to miss the point of the proposal. By requiring fundamental metaphysical theories to have exclusively ontological commitments, the truthmaker principle is clearly incompatible with my tense primitivism, all of whose commitments are ideological. This does not mean that we would have to reject the truism that truth supervenes on being; we merely have to spell it out in the correct way. According to my tense primitivism, truth supervenes on how things *were*, *are*, and *will be*, and that is all that can or needs to be said about this.

6

INSTANTS OF TIME

No THEORY of time is complete unless it includes an account of what instants are. My proposal is to adopt a 'linguistic ersatzism' about time that treat instants as maximal consistent sets of sentences of tense logic. This provides referents for ordinary dates like '16 June 1995' and for the more sophisticated temporal coordinates used in scientific theories. It also helps tense primitivists rebut the charge that tense logic has insufficient expressive resources to serve as a theory of time.

6.1 Definitional incompleteness

It is well known that there are finite sets of truth-functional sentence connectives that are definitionally complete in the sense that all other truth-functional connectives can be defined in terms of them. In particular, one can show that any truth-functional sentence connective of arbitrary adicity can be expressed in terms of our two connectives $\neg$ and $\rightarrow$. There is no comparable result for intensional operators. Neither the modal operator $\Diamond$ nor the tense operators P and F form a definitionally complete set for their respective classes of operators.

A familiar example in the modal case is the counterfactual conditional $\varphi \mathrel{\Box\!\!\rightarrow} \psi$ ('if φ were the case, ψ would obtain'). In the first rigorous treatment of this binary modal connective, Nelson Goodman (1955) tried to spell out $\mathrel{\Box\!\!\rightarrow}$ in terms of $\Diamond$ alone, but this project quickly ran into difficulties. Robert Stalnaker (1968) and David Lewis (1973) argue that any acceptable analysis of the counterfactual must appeal to a notion of comparative possibility that allows us to say whether one world is more possible than another. Roughly, the proposal is to count $\varphi \mathrel{\Box\!\!\rightarrow} \psi$ as true in a possible world just in case $\varphi \wedge \psi$ is more possible relative to that world than $\varphi \wedge \neg\psi$. To account for counterfactual conditionals in this way, we need to be able to say *how* possible a claim is, and $\Diamond$ only allows us to say whether it is possible at all.

The same problems arise for the binary tense operators S ('since') and U ('until'). A sentence like 'John was awake since dawn' is concerned with past times, but not with all of them. It asserts that John was awake at all past times that are later than the time of dawn. That is, $\mathsf{S}\varphi\psi$ is true at a time if and only if there is some earlier time at which φ is true such that ψ is true at all instants that are between these two times. The operator U works in a similar manner, just with regard to the future. Hans Kamp (1968) shows that S and U cannot be defined in terms of the monadic operators P and F because they only care about whether a time is past or future at all, and not whether it is more past or future than some other time.[38]

A second range of problems concerns operators that are two-dimensional in the sense defined by Krister Segerberg (1973). One example is the 'now' operator N, which is governed by the stipulation that $\mathsf{N}\varphi$ is true at a time if and only if φ is true at the present time.[39] Kamp (1971) proves that N can be eliminated in propositional tense logic, but claims that the quantificational case is different. More specifically, he argues that a quantified P/F tense logic is unable to express 'A child was born that will become ruler of the world'. If we follow Kamp in assuming a time-relative reading of the quantifiers then this sentence can be formalized with the help of the 'now' operator as $\mathsf{P}\exists x\,(Bx \wedge \mathsf{NF}Rx)$. The nested N ensures that Rx gets evaluated at times that are later than the present time, rather than at times that are later than the past time to which the leading operator P takes us. Kamp claims that there is no logically equivalent regimentation of this sentence in a tense logic that only has the tense operators P and F.[40]

A modal variant of this problem arises for the two-dimensional actuality operator A, which is defined by saying that $\mathsf{A}\varphi$ is true in a world if and only if the embedded sentence φ is true in the actual world. Allen Hazen (1978) proves that A is redundant in propositional modal logic, where sentences without modal operators suffice to express all claims about what is actually the case. The situation is different in quantified modal logic. Consider 'There could be something that does not actually exist'. If we use world-relative quantifiers whose range at a world comprises all the objects that exist there, then the actuality operator allows

[38]Meyer (2009b: sec. 8) gives a sketch of Kamp's proof. Some of the work of binary tense operators can be done by suitable two-dimensional ones. Gabbay (1977) presents a system with monadic two-dimensional tense operators in which S and U are redundant.

[39]Prior (1957b: 9) initially thought that this operator is redundant because its work is already done by the convention that the untensed sentences make claims about what is presently the case. He later changed his mind about this (1968a; 1968b).

[40]I say more about these issues in Meyer (2009a) and on p. 82 below.

us to formalize this as $\Diamond \exists x \, \mathsf{A} \neg \exists y \, y = x$, but the sentence seems to possess no regimentation in a quantified modal logic that has $\Diamond$ as its only modal primitive, and thus lacks the resources to refer to the actual world from within the scope of other modal operators.[41]

One might suggest that these problems merely show that we have made a poor choice of primitives, and that our modal theory of time merely needs to be formulated in terms of a larger set of tense operators. But while nothing prevents us from adding further operators, this does not seem to get rid of the underlying problem. Similar issues keep cropping up. Frank Vlach (1973) argues that P, F, and N are not definitionally complete, either, and that 'One day, all persons alive then would be dead' requires the introduction of a 'then' operator T. Johan van Benthem (1977) takes it one step further, by presenting a sentence that cannot even be formalized in terms of P, F, N, and T. There appears to be no finite set of tense operators that is definitionally complete.

Our tense logic even lacks a general way of expressing the claim that a sentence φ is true at *all* times. One could try to formalize this as $\mathsf{H}\varphi \wedge \varphi \wedge \mathsf{G}\varphi$, but this only works if the earlier-than relation is comparable. In a branching time series, it does not follow from φ's being true at all times that are past, present, or future relative to the present that it is true at all times. The sentence could still be false in some other branch. Unless there is a limit on the number of branches, φ is true at all times if and only if $\Theta\varphi$ holds for arbitrarily long sequences Θ of the tense operators H and G. If we expand our tense logic with additional axiom schemata to ensure comparability (p. 46) then we can embed the 'always' interpretation of S5 from p. 43 into the P/F calculus as a proper fragment, by defining $\Box\varphi$ as an abbreviation for $\mathsf{H}\varphi \wedge \varphi \wedge \mathsf{G}\varphi$. Otherwise, we need a separate 'always' operator.[42]

Since all of these problem cases are easily rendered in terms of explicit quantification over times, it seems that tense primitivism is unable to match the expressive resources of temporal substantivalism. Johan van Benthem (1977) and M. J. Cresswell (1990) conclude that we cannot account for time (and modality) in terms of primitive intensional

[41] See Hazen (1976), Crossley and Humberstone (1977), and Hodes (1984).

[42] Hughes and Cresswell (1975) discuss this issue in more detail. Thomason (1974; 1975) considers the converse question of whether the P/F tense logic can be embedded into S5. Of related interest is the argument by Bennett (1977) that an account of verb aspect requires the ability to talk about *time intervals*, which our simple tense logic cannot do, either. See also Kamp (1979) and Kuhn (2002). A tense logic for time intervals is developed in Humberstone (1978), van Benthem (1980), and Burgess (1982).

operators. However, the interpretation of these inexpressibility results is not quite that straightforward. When considering the definitional completeness of truth-functional connectives, we are asking whether any such connective can be expressed in terms of certain other connectives of the same type. It is obvious that this is not true for tense operators. Not even H and G can be defined in terms of P and F unless we also have negation at our disposal. The interesting question about tense operators is whether we can express them in terms of our primitive operators *plus* the resources of a suitably chosen background logic. Any expressive shortcomings could therefore be due to an insufficient background logic, rather than to a defect of tense primitivism.

I want to argue that tense primitivists can compensate for the limitations of P and F by acquiring the ability to quantify over times themselves. Instead of adopting the substantivalists' view that instants of times are metaphysically basic entities, they can treat them as derived entities that are constructed from something else. Earlier, we said that the models of tense logic are composed of 'times', but this terminology must not be taken too literally. The aim of a model theory is to characterize the theorems of our logic, and for that purpose the precise nature of the elements of the models is largely irrelevant. All the work in the proof of the soundness and completeness theorem is done by the restrictions the clauses on p. 45 impose on $\models$. There might also be an intended model that is made up of *real* times, but that would at best allow us to characterize tense operators in terms of the times in this intended model. What is at issue here is the exact converse. The task is to explain what times are in terms of tense operators, not the other way around.

6.2 Linguistic ersatzism about times

The account of times I favour is a tense version of what Lewis (1986c) calls *linguistic ersatzism* about possible worlds.[43] As developed by Andrew Roper (1982), the linguistic ersatzer identifies possible worlds with maximal consistent sets of sentences of modal logic. A set of sentences is *maximal* if it contains either φ or $\neg\varphi$ for every sentence φ of our language. Given a system of modal logic, such a set is *consistent* if it does not contain sentences $\psi_1, \ldots, \psi_k$ such that $\vdash \neg(\psi_1 \wedge \ldots \wedge \psi_k)$, where $\vdash$ denotes derivability in the modal system at issue. Suppose we

[43] Proponents of the view do not think that there is anything 'ersatz' about their possible worlds; they believe their worlds are the real thing, rather than an inferior substitute. But since the use of the term 'ersatzism' has become well established, I shall adopt it here.

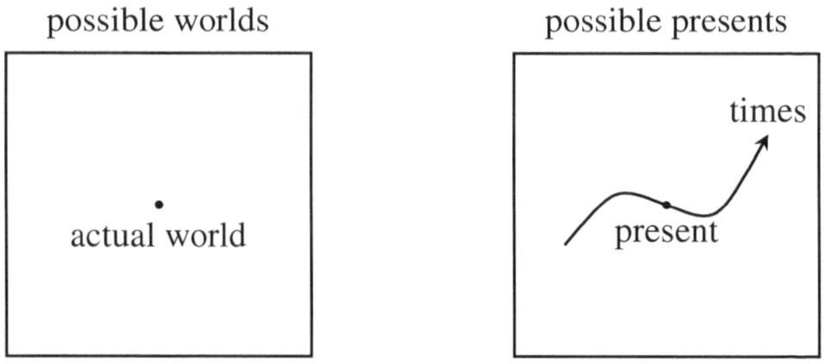

Figure 6.1. Possible worlds and possible presents

say that a sentence φ is *true at* a set of sentences s if and only if $\varphi \in s$, which we write as $s \mid \varphi$.[44] Then a maximal consistent set of sentences is a way of assigning truth values to all sentences that is compatible with what our modal logic tells us about the logical relations amongst them.

The temporal case is slightly different. Suppose we call a maximal Z-consistent sets of sentences of tense logic a *possible present*. Then not all possible presents qualify as times. At best, times are those possible presents that did, do, or will happen, and most of them never do. If we adopt the system S5 from p. 42 as our theory of modality then everything that is possible is *necessarily* possible (M4), which means that what possible worlds there are does not depend on which of them is actual. But if we read M4 as an axiom of tense logic then it only claims that what is sometimes true is *always* sometimes true, not that it is *necessarily* sometimes true. The same possible present can therefore qualify as a time in one possible world, but not in another. This is a key difference between time and modality. Since it is a contingent matter which possible presents are times, the construction of the time series requires one more step than the construction of the set of all possible worlds. After we have obtained possible presents as maximal consistent sets of sentences of tense logic, we still need to determine which of them are times, and how they are arranged by the earlier-than relation.

Here is how this can be done. We start by picking a maximal Z-consistent set of sentences p as our choice of present time. Then p not only purports to describe how things presently are, it also contains implicit descriptions of all other times. These descriptions are hidden within the scope of the tense operators that occur in the sentences in p.

[44]The operator | is due to Myro (1986a;b) and works like the R_t of Rescher (1966).

To extract this information, we define earlier-than ($\lhd$) and later-than ($\rhd$) relations on sets of sentences:

$$s \lhd s' \text{ iff } s' \mid \mathsf{P}\varphi \text{ whenever } s \mid \varphi$$
$$s \rhd s' \text{ iff } s' \mid \mathsf{F}\varphi \text{ whenever } s \mid \varphi$$

For any tense logic that contains axiom Z1 and the temporal generalization rule TG, we can prove that $\lhd$ and $\rhd$ interact with the tense operators as expected. If a possible present p contains some past (future) tense claim then there is a possible present that is earlier (later) than p at which the embedded sentence is true.[45]

$$(*) \quad \begin{array}{l} p \mid \mathsf{P}\varphi \text{ iff there is a possible present } t \text{ such that } t \rhd p \text{ and } t \mid \varphi \\ p \mid \mathsf{F}\varphi \text{ iff there is a possible present } t \text{ such that } t \lhd p \text{ and } t \mid \varphi \end{array}$$

If we also have axiom Z2 then the relations $\lhd$ and $\rhd$ are easily shown to be converses of one another.

In light of these results, one might now propose that a possible present counts as a time if and only if it is either earlier than, later than, or identical to our choice of present p. A drawback of such a definition is that it would arbitrarily rule out the possibility of branching time series. If we admit branches then there can be a time that is future relative to some past time, but is itself neither past, present, or future relative to the present time because it is located in some other branch of time. There are stronger systems of tense logic that prohibit branching, but Z does not. Consider the following possible present:

$$p = \{\mathsf{H}\varphi, \varphi, \mathsf{G}\varphi, \mathsf{PFG}\neg\varphi, \ldots\}$$

In this case, there is a possible present t_1 earlier than p at which φ is true, and in whose future there are two branches: one that contains p and in which φ is always true and another branch that contains at least one other possible present t_2 and in which φ is always false (Fig. 6.2).

Since I do not want to rule out branches by fiat, let me define the *ersatz time series* T_p with present p as the smallest set of possible presents that contains p and is closed under the earlier-than and later-than relations. To construct such a time series, we start out with our choice of present p and then add all possible presents that are either earlier or later

[45]The proof can be found in Meyer (2009b: prop. 3.1) and extends to any system of tense logic that is stronger than the simple system Z.

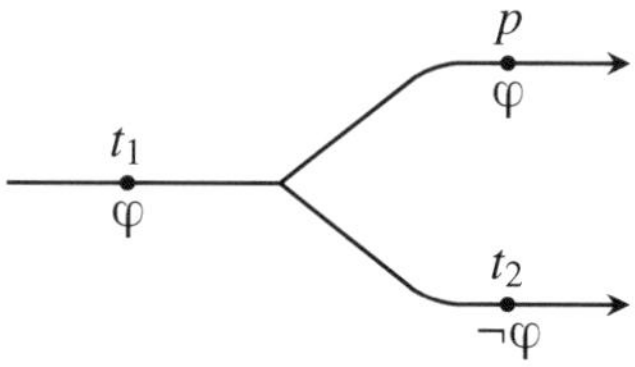

Figure 6.2. A branching time series

than p. In the next step, we add all possible presents that are earlier or later than some possible present we added in the first step, and so on. To put this more rigorously, say that a finite string of possible presents $\pi_0, \pi_1, \ldots, \pi_n$ is *connected* just in case either $\pi_0 = \pi_n$, $\pi_{i+1} \lhd \pi_i$, or $\pi_{i+1} \rhd \pi_i$ for all $0 \leq i < n$. Then a possible present t is an element of T_p just in case it can be reached from p by a connected string of possible presents. A sentence φ is said to be true in the time series T_p if and only if it is true at its present time, $p \mid \varphi$.

Before returning to the question of expressive power, let me make four quick remarks about this construction. First, the property-time link is built into the very foundation of this account of instants. To attribute a property K to an object a, we have to include either $\mathsf{P}Ka$, Ka, or $\mathsf{F}Ka$ in our choice of present p. With the key result $(*)$, any one of these three options entails the existence of an ersatz time t such that $t \mid Ka$. So we cannot attribute a property to an object without committing ourselves to the existence of a time at which it possesses that feature.

Second, different choices of p yield different time series. Which of them produces the actual time series depends on which possible present correctly describes how things presently are, but it is important to note that the time series does not depend on which *time* is present. Suppose we pick an arbitrary time $q \in T_p$. By taking the minimal closure of q under $\lhd$ and $\rhd$, we then recover the same time series as the one we started out with; that is, $T_q = T_p$. Any instant in a given time series thus contains a complete description of all other times, and the passage of time does not change what times there are.

Third, since we are unlikely to know the truth values of all the sentences of our language, we might not be able to execute the construction of the ersatz time series. But we do not need to be temporally omniscient to *refer* to times. Ersatz times are often determined by a single element. If temporal coordinate systems work as advertised then we can pick out times by the position of a few physical objects, such as the sun, the moon, and the hands on a clock. This yields a range of relatively short *calendar sentences* that are true at one time only, and which allow us to refer to times by means of definite descriptions. It is not necessary that there are such calendar sentences, but our world seems to cooperate.[46]

Fourth, even though time series are not the same as *models* for tense logic, the two are intimately related. Given a model $\langle \mathfrak{T}, <, \mathfrak{p} \rangle$ of tense logic, the set of sentences $\{\varphi : \mathfrak{p} \models \varphi\}$ is maximal Z-consistent and all sentences take the same truth values in the model $\langle \mathfrak{T}, <, \mathfrak{p} \rangle$ as they do in the time series $T_{\{\varphi : \mathfrak{p} \models \varphi\}}$. This means that any sentence that is true in all time series is true in all models. Conversely, any maximal Z-consistent set of sentences p defines a model $\langle T_p, \lhd, p \rangle$ of tense logic in which all and only the elements of p are true. Since any sentence that is true in all models of tense logic is true in all time series, time series can thus perform all the logical work of models in the proof of the soundness and completeness theorem: a sentence is a theorem of the tense logic Z if and only if it is true in all ersatz time series.[47]

6.3 Times in the object language

The construction of times as maximal consistent sets of sentences takes place in the meta-language of tense logic, but one can easily provide the object language with the ability to talk about these times. Suppose we take the propositional tense logic from Section 4.3 and add the lower-case letters p, q, and r as *dates*, which are singular terms for times. Numerals may be added as subscripts to ensure an inexhaustible supply of dates. In interpreting this language, dates refer to times, which are maximally consistent sets of sentences of the original language of tense logic (before the introduction of dates). If α is a date and φ a sentence of the original tense logic then we write $\alpha \mid \varphi$ to say that φ is an element of the set of sentences picked out by the date α.

Because times are maximal consistent sets of tense logic, the following are then theorem schemata of this extended tense logic:

$$\neg\alpha \mid \varphi \leftrightarrow \alpha \mid \neg\varphi \qquad \alpha \mid (\varphi \rightarrow \psi) \rightarrow (\alpha \mid \varphi \rightarrow \alpha \mid \psi)$$

[46]The issue of naming times and possible worlds is also discussed in Blackburn (1993), Gabbay and Malod (2002), and Cresswell (2010).

[47]This result extends to the stronger systems of tense logic discussed on p. 45. The proof is tedious and I shall omit it here; see Meyer (2009b: sec. 9) for details.

It is convenient to reserve one of our dates as a name for the present time of our time-series. Suppose we use the letter p for this purpose. This means that we use the same letter as a name for the present in the object language as we did for the present in the meta-language. With the referent of p fixed in this way, we get another theorem schema:

$$p\,|\,\varphi \leftrightarrow \varphi$$

Prefixing the operator $p\,|\ldots$ is therefore always redundant.

If α and β are both dates then we write $\alpha \lhd \beta$ just in case $\beta\,|\,\mathsf{P}\varphi$ for all φ such that $\alpha\,|\,\varphi$. The later-than relation $\rhd$ gets defined in terms of $|$ and F in an analogous way. To endow our language with the ability to quantify over ersatz times, we also add identity, time quantifiers $\exists$ and $\forall$, and time variables s, t, and u (with or without numerals as subscripts). If φ is a sentence of the underlying tense logic and α and β are dates then $\alpha\,|\,\varphi$, $\alpha \lhd \beta$, $\alpha = \beta$, $\alpha \rhd \beta$, and φ itself are sentences of the extended language. Truth-functional compounds of sentences of the extended language are also sentences, as is any expression that can be obtained from a sentence by using time quantifiers and variables to quantify into date positions. Nothing else is a sentence.[48]

These additions provide us with a few more theorem schemata for our extended object language. Because our tense logic contains axiom Z2, the relation $\lhd$ is easily shown to be the converse of $\rhd$:

$$\forall s \forall t (s \lhd t \leftrightarrow t \rhd s)$$

There is also an object-language version of the result reported on p. 61, which shows that quantification over times, tense operators, and the earlier-than and later-than relations interact as expected:

$$\alpha\,|\,\mathsf{P}\varphi \leftrightarrow \exists t (t \lhd \alpha \wedge t\,|\,\varphi) \qquad \alpha\,|\,\mathsf{F}\varphi \leftrightarrow \exists t (t \rhd \alpha \wedge t\,|\,\varphi)$$

The Indiscernibility of Identicals becomes the theorem schema:

$$a = b \rightarrow \forall t\,(t\,|\,Ka \leftrightarrow t\,|\,Kb)$$

Tense primitivists could thus solve the Problem of Change by saying that an object a has a property K at a time t if and only if $t\,|\,Ka$.

[48] These stipulations prohibit us from iterating expressions of the form $\alpha\,|\ldots$ No such operator is allowed to occur within the scope of another and expressions like $\alpha\,|\,\beta\,|\,\varphi$ do not count as sentences. This simplifies the exposition of our language and nothing of substance is being lost by these restrictions. Since times are sets of abstracta, they have the same elements at all times and there are no interesting temporal facts about them that our language needs to describe. There is no reason to scratch where it does not itch.

This extra structure allows the object language to talk about times and what is true at them. Since it cannot express the notion of a maximally consistent set of sentences of tense logic, it does not permit the object language to say what times are. But the extra structure is sufficient to rebut the objection that tense primitivism lacks the expressive resources to serve as a theory of time. Whenever their opponents use quantification over times to explain some temporal notion, tense primitivists would do exactly the same. The only difference is that their temporal variables range over ersatz times, rather than metaphysically basic instants on a manifold $\mathcal{T}$.

This strategy does not yield a sentence-by-sentence translation of every temporal claim into the original language of tense logic, and therefore does not refute the inexpressibility results reported above. In general, we do not get a regimentation that only contains P and F, and no quantification over ersatz times. But such a regimentation is not needed, anyway. What is important is that the additional structure that we introduced does not expand the *temporal* ideology or ontology of our theory of time. Given that two maximal consistent sets of sentences of tense logic only differ if they disagree about the truth value of some sentence of the underlying tense logic, tense primitivists would still be entitled to their claim that P and F capture everything that there is to be said about time. The expressive limitations of the tense logic Z would only be a problem for tense primitivism if they concerned recognizably temporal matters, and they do not.

There is another way of adding quantification over times to a system of tense logic. Instead of treating them as maximal consistent sets of *sentences*, Kit Fine suggests that times are maximal *propositions*.[49] This proposal enriches the object language with modal operators and propositional quantifiers that range over subsets of the set of possible worlds used to interpret the modal operators. Two propositions are thus counted as identical if and only if they are true in the same worlds. A proposition A is said to be maximal just in case it is possible and would entail all other true propositions if it were true:

$$\Diamond A \wedge \Box(A \rightarrow \forall X(X \rightarrow \Box(A \rightarrow X)))$$

[49]Fine (2005b); see also Prior (1967: ch. V). Zalta (1988: ch. 4) argues that times are determined by the propositions true at them, but are not themselves propositions. Propositional quantifiers are introduced in Fine (1970), Bull (1969), and Kaplan (1970).

Given two maximal propositions A and B, we can say that the former is earlier than the latter, $A \lhd B$, if and only if $\Box(A \rightarrow \mathsf{F}B)$. Similarly, we can define $A \rhd B$ as shorthand for $\Box(A \rightarrow \mathsf{P}B)$. The time series derived from a maximal proposition P chosen as present would then be the smallest set of maximal propositions that contains P and is closed under these earlier-than and later-than relations.

This is an alternative to the times-as-sets view, but it makes substantial assumptions about what propositions there are. Fine's proposal can only deliver more than one time if at least some propositions are temporal and change their truth values over time. Suppose P is true now and that $P \lhd A$. By definition of $\lhd$, this yields $\Box(P \rightarrow \mathsf{F}A)$ and thus $P \rightarrow \mathsf{F}A$ with the S5 axiom M2 from p. 42. Hence $\mathsf{F}A$ by modus ponens. So if P and $P \lhd A$ are true then A *will* be true. Can A be true now as well? Suppose it were. Then we would get $\Box(P \rightarrow A)$ because P is a true maximal proposition. But A is itself a true maximal proposition, which yields $\Box(A \rightarrow P)$ because P is true now. So P and A are necessarily equivalent and thus identical on our account of propositions. Hence there can only be a time A other than P such that $P \lhd A$ if the proposition A is now false, but will be true, which requires A to be temporal.[50]

Since maximal propositions are vastly more specific than anything that can be expressed by the utterance of an actual sentence, temporal discourse would therefore be partly about a language-independent realm of temporal propositions. The advantage of the times-as-sets view is that it preserves the ontological frugality of tense primitivism. It provides the needed referents for dates without incurring commitment to temporal propositions, temporal parts, metaphysically basic time points, or any other recognizably temporal entities.

[50]Bourne (2006: sec. 2.V) defends an account that treats times as maximally consistent sets of *untensed* propositions, and which shifts all the information about the order of the time series into an un-analysed earlier-than relation.

7

TEMPORAL STRUCTURE

WHILE temporal substantivalists think that the manifold $\mathfrak{T}$ has its structure independently of what temporal processes happen to unfold, linguistic ersatzism about time is committed to the opposite position. If times are sets of sentences then the structure of the time series is not only a contingent matter, it is also dependent on the expressive resources of our language. More needs to be said to establish that this is indeed a plausible view about temporal structure.

7.1 The number of times

In the modal case, W. V. Quine (1969) objects that linguistic ersatzism cannot deliver enough possible worlds to give an acceptable account of possibility. Since there are as many sentences in our countable language as there are natural numbers, $\aleph_0$ many, Cantor's Theorem entails that there are as many sets of sentences as there are points in the continuum, $\aleph_1$ many. That is a fairly large number, but there are even more possibilities. Suppose we are given a spatial manifold with continuum many spatial points. If we grant that all of these points could either be occupied by matter or not then there are at least as many possible distributions of matter as there are subsets of the continuum, of which there are $\aleph_2$ many. Hence there are fewer ersatz worlds (maximal consistent sets of sentences) than there are possibilities.

This is not a problem in the temporal case because nobody believes there to be more than continuum many times. Moreover, we can easily prove that linguistic ersatzism can deliver time-series with cardinality $\aleph_1$. Consider an object that moves on a curve $\mathscr{C}$ on a continuous manifold (see Fig. 3.6 on p. 35). If we adopt a formal language that allows us to describe this motion then we can distinguish times by the successive location of the object. Since the object passes through continuum many spatial points, this yields $\aleph_1$ many different ersatz times. Without abandoning the assumption that our formal language has a countable vocabulary, we cannot just add names for *all* the points on the curve $\mathscr{C}$, but this is not needed, anyway. It suffices to have names for a countable dense subset $\mathscr{C}^*$ of points on the curve. Similar to Richard Dedekind's construction of real numbers as sets of rational numbers, we can then identify any point on the curve with the subset of points in $\mathscr{C}^*$ that the object passed to get to that position.[51] Since this proof is based on the assumption that there is a spatial manifold with continuum many points, which may or may not be the case, this does not show that one can get continuously many times in every possible world. Linguistic ersatzism openly admits the possibility that the time series might be countable, discrete, or finite. There are even maximal Z-consistent sets that contain neither $\mathsf{P}\varphi$ nor $\mathsf{F}\varphi$ for any sentence φ. In that case, there are no possible presents that are earlier or later than p, and the ersatz time series would consist of a single instant.

The number of ersatz times often depends on how many changes occur, but we can also get *more* times than changes. Linguistic ersatzism admits the possibility of *freezes*, which are extended periods during which no object undergoes change.[52] Suppose that genuine change consists in the acquisition or loss of a natural property, and suppose further that all such properties are picked out by the primitive predicates of our language. Then the possible present $\{Ka, \mathsf{P}Ka, \mathsf{PP}Ka, \mathsf{PPP}Ka, \ldots\}$ yields an ersatz time series with four instants at which Ka is true, even if no other atomic sentences change their truth values during this period. Since our predicates are assumed to exhaust all natural properties, this gives us four different times at which no object undergoes change.

What ersatz times we get partially depends on the expressive resources of our language. By choosing more fine-grained predicates that allow us to describe the world in more detail, we might get more ersatz times than we would obtain otherwise. This is one issue on which linguistic ersatzism interestingly diverges from relationist accounts of time. Both views regard times as certain abstraction, but if times are sets of temporally ordered events then their number only depends on what events happen to occur, not on the linguistic resources we adopt for describing them. Like temporal substantivalists, relationists believe that there is an objective fact of the matter how many times there are.

Tense primitivists could have a similar kind of account, by adopting Kit Fine's view that times are maximal propositions from p. 65.

[51] See Meyer (2009b: sec. 5) for more details.

[52] Those who are convinced by Shoemaker's (1969) argument that we could have empirical evidence for freezes should welcome this result. See also Schlesinger (1970), Earman (1970: sec. III), and Warmbrōd (2005).

Since maximal propositions are language-independent entities (there is no actual sentence that expresses them), the structure of a time series made up of maximal propositions would be independent of the expressive resources of our language. Some might welcome this feature, but this view would not really give us a much better answer to the question of how many times there are. To find the number of times we would do the same thing a linguistic ersatzer would do: find as many true tensed sentences as we can and then figure out how many maximal propositions, or maximal consistent sets of sentences, we need to accommodate them all. The advantage of the times-as-sets view is that it gives tense primitivists as many times as they could possibly want, up to $\aleph_1$ many, without committing them to any view about propositions.

7.2 Eternal recurrence

Locally, there is no unacceptable limit on the number of ersatz times. The global structure of the time series is a different matter. Temporal substantivalists can distinguish two different types of eternal recurrence: *linear* and *circular* recurrence.[53] In linear recurrence, the temporal manifold $\mathcal{T}$ looks like a straight line and infinitely many qualitatively indistinguishable episodes H succeed one another in a linear fashion. In a case of circular recurrence, $\mathcal{T}$ forms a loop, with one episode H stretched out along its circumference.

Linguistic ersatzers about time cannot make this distinction. If we put the present p at the same place within an episode H, then the same sentences of our tense logic are true in both cases. But since ersatz times inherit their identity conditions from set theory, there cannot be two times at which the same sentences of tense logic are true. If everything that our language can describe repeats itself after a fixed period then the resulting ersatz time series is perforce circular, which means that linguistic ersatzism is unable to account for linear recurrence.[54]

Comparable remarks apply to *symmetric branches*. Temporal substantivalists can distinguish a linear time series from one that has two exactly similar future-directed branches. Linguistic ersatzers must reject this possibility; their view only permits asymmetric branches in which

[53] Earman (1995: ch. 7) discusses the physics of eternal recurrence.

[54] There are *models* of tense logic that describe linear recurrence, but no ersatz time series. We noted in Section 6.2 that any model yields a time series in which all sentences take the same truth values, but there is no guarantee that this time series is isomorphic to the model. The two non-isomorphic models in Fig. 7.1 give rise to the same time series.

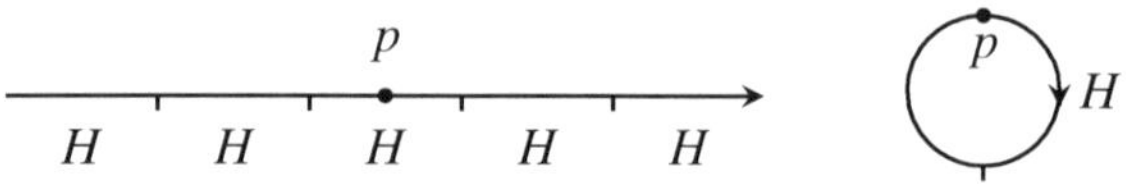

Figure 7.1. Linear and circular recurrence

something is true in one branch that is false in the other. While temporal substantivalism permits time to have a certain amount of 'superfluous' structure, linguistic ersatzism always yields the smallest time series that accounts for all true tensed claims.

These are noteworthy differences, but I do not think they count against linguistic ersatzism. Temporal substantivalists can distinguish linear from circular recurrence because they already believe the number of times to be independent of what is happening within time. That is precisely what linguistic ersatzers deny. Given their view about what times are, linguistic ersatzers should simply reject the possibility of linear recurrence. Linguistic ersatzism is not meant to describe a language-independent manifold of time points; all that matters is that the account captures what tense primitivists think there is to be said about time.

The case of eternal recurrence also helps to bring out another issue for linguistic ersatzism. When discussing the logic of tense operators in Section 4.3, we identified axioms for P and F with specific structural properties of the earlier-than relation. One property that was notably absent from this list is *antisymmetry*. If we define a transitive earlier-than in the obvious way then this relation is antisymmetric in the case of linear recurrence but not in the case of circular recurrence. Since linguistic ersatzism does not admit linear recurrence, it could only guarantee an antisymmetric and transitive earlier-than relation if it managed to prohibit eternal recurrence altogether.

One could achieve this by postulating the existence of a first or last time, but that would also rule out all models in which the time series looks like the real numbers. To ensure non-recurrence in a way that permits a time series without beginning or end, we would need assurance that the future is always different from the past, and that cannot be guaranteed with an axiom schema. Schemata can do some of the work of universal quantification over sentences, but they cannot be used to formulate the existential claim that there is some sentence that will be true, relative to the time at issue, but never was true in the past.

If we add time quantifiers and $\lhd$ to the object language of our tense logic, as described in Section 6.3, then we could characterize the antisymmetry of earlier-than as $\forall s \forall t(s \lhd t \rightarrow \neg t \lhd s)$. But *saying* that $\lhd$ has this property is not the same as presenting a principle that ensures that it does. This would be a problem if the antisymmetry of earlier-than were a necessary truth that needs to be expressed by our fundamental theory of time, and to me the very possibility of circular recurrence suggests otherwise. Any inclination to think that earlier-than is necessarily antisymmetric is probably just due to the salient fact that our experience of the actual world is not one of recurring events. Even though the future is always different from the past, it seems perfectly coherent to suppose that the truth values of tensed sentences are distributed in such a way that the resulting ersatz time series is circular. If that is right, then the antisymmetry of the earlier-than relation only holds contingently.[55]

The same seems to hold for the transitivity of earlier-than. If earlier-than were transitive in the case of circular recurrence then every time would be earlier than every other one. Nothing would distinguish points on one side of p from those on the other, and there would be no real substance to the claim that the time series looks like a *circle*. Bas van Fraassen (1970: sec. 3.1) and William Newton-Smith (1980: sec. 3.2) conclude that circular time cannot be described in terms of a binary earlier-than. But this only follows if we assume that the relation is transitive. Mark Reynolds (1994) presents a tense logic that characterizes circular time by giving up transitivity. He thinks of the future as stretching ahead, halfway around the circle, and the past as stretching halfway back so that it just meets the future on the other side. If this is a coherent view then the transitivity of earlier-than is also a contingent matter, and need not be written into our fundamental theory of time, either.

It was partially for these reasons that Section 4.3 resisted the temptation to expand our tense logic beyond the simple system Z. While one could easily add a number of further axiom schemata to this minimal system, a linguistic ersatzer about time would not regard any of the corresponding structural properties as necessary features of the time series.

7.3 Objective distances

By defining the earlier-than $\lhd$ relation in terms of primitive tense operators, as described on p. 61, linguistic ersatzers can give an account of

[55] Pap (1962: 97) claims that 'no event precedes itself' is an example of a synthetic a priori truth, but he does not give any argument in support of this claim.

the order in which times succeed one another. It is less obvious whether they can also give an acceptable account of the amount of time that has elapsed between two instants. One could of course define a temporal coordinate system that assigns to each ersatz time a number as its date, and then assess the temporal distance between two instants in terms of the numerical difference between their respective coordinates. But this falls short of what is needed. Especially for applications in physics, we need an account of objective durations, and a definition that arbitrarily stipulates that one set of sentences is a certain temporal distance from another set is not going to give us that. This is a pressing issue, but it is no less urgent for spatial accounts of time. In a discrete time series we can determine the temporal distance between two instants by counting the number of intervening times. The duration of a time interval would be the number of instants it contains. If the time series forms a continuum, though, then any theory of the nature of time has to find a way of dealing with the fact that there is always the same number of time points between any two arbitrarily chosen instants, namely $\aleph_1$ many. As Adolf Grünbaum (1968) puts it, the continuum is 'metrically amorphous'.

My proposal is that linguistic ersatzers give the same account as everybody else, by defining durations in terms of appropriate measurement procedures. One measures the duration of a process by comparing it to another process. Sometimes the reference process is chemical or biological, like the growth of tree rings or the decay of carbon isotopes, but more often we measure durations by measuring distances. As suggested by Leonhard Euler (1748), we call two durations equal if a uniformly moving object covers the same distance during both durations. We first pick an object that is assumed to move at constant velocity, such as a light ray in a vacuum, and then determine how far this reference object has travelled between the beginning and the end of the process to be measured. Dividing this distance by the velocity of the reference object yields the time elapsed. A variant of this method uses a periodic motion, such as a swinging pendulum or a planetary orbit, and determines durations by counting the revolutions of the reference object during the time interval in question. In this way, days are measured by rotations of the earth around its axis, lunar months by rotations of the moon around the earth, and solar years by rotations of the earth around the sun.

Ernst Mach objects that this would make our account of duration unacceptably circular because the notion of uniform motion already assumes the notion of duration that we are trying to define:

> If a thing *A* changes with time then this only means that the circumstances of *A* depend on the circumstances of another thing *B*. ... We are entirely incapable of measuring the changes of things against *time itself*. Time is an abstraction that we arrive at through the changes of things. ... One motion can be uniform relative to another. The question of whether a motion *in itself* is uniform *has no sense at all*. Nor can we speak of 'absolute time' (independently of any change).
> (Mach 1912: sec. 2.6)

To judge the motion of an object *A* to be uniform, we need to compare it to the motion of a second object *B*. To judge *B*'s motion to be uniform we need to compare it to that of a third object *C*, and so on. This either ends up being circular, if a reference object occurs more than once in this series, or else leads to an infinite regress in which we never succeed in obtaining the desired definition of duration.

I think this is a pseudo-problem. What is at issue here is a general feature of theoretical terms, not something that is peculiar to durations. By way of illustration, consider the task of defining properties of mass. An obvious suggestion is that such terms get defined operationally, in terms of our procedures for measuring mass. For instance, one could say that an object has mass one kilogramme if and only if it balances a pair of scales with the standard kilogramme. This might look appealing at first, but it quickly runs into the problem that any given measurement procedure is of limited scope. The standard kilogramme method does not allow us to determine the mass of an electron, the mass of the moon, or even the masses of medium-sized objects outside France, where the standard kilogramme is located. If they are to be attributed any mass at all, these other objects require different measurement procedures.

At best, attributes of mass are pinned down by a cluster of measurement procedures whose ranges of application jointly cover all cases. This raises the question of whether a proposed cluster definition is consistent, and only a scientific theory about properties of mass can guarantee that the different measurement procedures yield the same result in those cases where their ranges of application overlap. This does not mean that theoretical terms cannot be defined in terms of measurement procedures, but it does show that operational definitions cannot proceed in isolation from our theories about the quantities in question. Theoretical terms must be defined implicitly by the scientific theories in which they appear. Mass is that property that does what our best theories of motion say mass does: it makes an object attract other massive objects, causes it to accelerate in a certain way when acted upon by forces, and so on. As Carl Hempel puts it, 'In scientific inquiry, concept formation and theory formation must go hand in hand'.[56]

Similar remarks apply to the task of defining spatial and temporal coordinate systems. Any method for determining durations and distances is of limited scope. The pendulum method does not allow us to determine the duration of the last ice age, and carbon dating is of no use for measuring the time it takes for a light ray to travel from the computer screen to my eye. Likewise, comparing distances to the standard metre bar does not help to determine the distance between Venus and Mars, or the distance between two electrons. In all but the most trivial cases, we need a cluster of measurement procedures to assign each instant of time a real number as temporal coordinate, and each point in space a triple of real numbers as spatial coordinate. Some aspects of coordinate systems, such as choice of units of length and duration, are purely conventional, but only a theory of the spatial and temporal processes involved can ensure that such a cluster definition of coordinates is consistent. Our physical theories help to define the spatial and temporal coordinates in terms of which they themselves are formulated.

Mach thus overlooks the possibility that we can have theoretical reasons for judging certain motions to be uniform. The reason for thinking that light in a vacuum moves at a constant speed, or that a pendulum performs a periodic motion, is that Maxwell's equations of electrodynamics and Newton's laws of mechanics tell us that they do. Without a theory to provide standards of uniformity, it makes no sense to ask whether an object's motion is really uniform, just as it makes no sense to ask, without reference to a theory of mass, whether one object really has twenty million times the mass of another. Newton's and Maxwell's theories are themselves formulated in terms of the notion of duration that we would define with their help, but this does not make the definition of uniformity circular; it is a general feature of theoretical terms.

However, this does leave us with an interesting question. Suppose we accept that temporal distances are implicitly defined by our best theories of motion, and suppose further that some such theory is in fact true. What does this tell us about the make-up of reality? One option is to say that, if the theory is true, there must be theory-independent temporal distance relations that the theory correctly represents. For lack of a better term, let me call this view *realism* about temporal distances. It

[56]Hempel (1966: 97); see also Hempel (1958) and Lewis (1983a).

is not clear that we must be realists in this sense, but if we are then we need appropriate relata for these temporal relations.

Spatial views of time have no problem with this. Temporal substantivalists could postulate distance relations between the points on the manifold $\mathcal{T}$ and relationists about time could introduce them between events or temporal parts. Either way, temporal distances would be *external* relations. Since the points on $\mathcal{T}$ are qualitatively indistinguishable, the distances between them cannot be determined by the intrinsic nature of these points alone. Temporal distances would have to consist of additional relations that are independent of the other features of the temporal manifold. Relationists about time could perhaps account for the durations of larger events in terms of their intrinsic features, by appealing to the temporal distances between smaller events, but ultimately they also have to postulate primitive temporal distance relations that do not supervene on the intrinsic properties of the events they relate.

Tense primitivists cannot be realists about temporal distances because they do not postulate any metaphysically basic temporal objects that could serve as the relata of such relations. If times are sets of tensed sentences then any temporal relations between them would have to be determined by their elements, and sentences do not stand in distance relations to one another. What tense primitivists should say instead, I propose, is that standards of temporal distance are always theory-relative, even if the theory is true. The immediate motivation for this view is that it is forced upon us by tense primitivism, but there are independent reasons for rejecting realism about both spatial and temporal distances.

Henri Poincaré (1902: ch. 4) presents the example of a world whose geometry can be described in two radically different ways. One possibility is to think of it as the inside of a large finite sphere of radius R that is filled with a variable temperature field. Inside the sphere, the temperature (in Kelvin) is proportional to $R^2 - r^2$, where r is the distance from the centre. The world also has a variable refractive index that is proportional to $(R^2 - r^2)^{-1}$, which makes light rays move on circular trajectories, rather than on straight lines. Since the temperature is highest in the middle of the sphere, and gradually declines to absolute zero on the boundary, the measuring rods we use to define spatial coordinates shrink as we move radially away from the centre, making equidistant points appear farther away from each other. The second way of thinking about the world is to regard it as an infinite space with a non-Euclidean geometry. We start by stipulating that light rays move on straight lines and adjust the geometry accordingly. Given the way the example is set up, no experiment can distinguish between the two characterizations of the geometry of Poincaré's world, even though they make different claims about spatial distances. A similar example can be constructed in the temporal case. Consider a world that can be regarded as possessing either a finite future (without a last time), or an infinite future in which all physical processes (including clocks) asymptotically come to a standstill. If we are clever in postulating the physical features of this world then there is again no experiment that could distinguish between the two characterizations of temporal distances.

Realists might argue that these are just instances of the familiar underdetermination of theory by data, and that a theory's secondary virtues, notably its overall simplicity, can be used to decide between these competing views about spatial and temporal distances. But what is at issue here is underdetermination by *all possible* evidence, not merely underdetermination by the data that has actually been collected. Perhaps it is rational to adopt the simplest account of the available evidence, but it does not follow from this that one of Poincaré's two readings must be true just because it is simpler. Both scenarios seem equally possible and we surely cannot rule out a priori that the world is not a messy place. If we insist on realism about distances and durations then we are bound to end up with a view on which the structure we postulate is entirely beyond our ken, which is reason not to postulate such structure at all.

One might be tempted by the view that, as a matter of *metaphysics*, there must be some real spatial and temporal structures that underwrite our theories of motion, even if our *physics* is unable to determine the precise details of this structure. But this assumes that there is a sense of 'distance' and 'duration' that is independent of our best physical theories, and we just saw that this is in fact false. Since coordinate systems are partially defined by the scientific theories that use them, it would be utterly unclear what would make it the case that some relations of distance and duration are the 'real' ones. These relations cannot be singled out by the mere fact that there are *true* theories of motion about them because Poincaré's example shows that there are many other geometries and chronometries that satisfy the same condition.

These considerations count against realism about distances and durations, but they do not show that geometry and chronometry are merely a matter of convention. Scientific theories still impose a significant constraint on available coordinate systems. Consider the case of forces and acceleration discussed on p. 30. We can only talk about an object's acceleration with reference to a particular spatial and temporal coordinate system. This means that also our account of forces, which bring about accelerations, depends on our choice of coordinates. On the other hand, we have to appeal to dynamical theories, which are theories of forces, to define such coordinate systems in the first place. What we need is therefore a combination of temporal and spatial coordinate systems and a theory of motion that is empirically successful. This is not something that can be obtained by mere stipulation. Once we have found such a combination, it is easy to generate others, by compensating any 'distortion' of the geometry or chronometry with a corresponding modification of our physical theories. The choice amongst such empirically indistinguishable accounts is then a pragmatic matter, and we are free to pick an account that strikes us as simple. But that neither entails that this must be the true account, as realists about distances might claim, nor does it change the fact that, relative to a fixed account of a theory of motion, there are objective facts about distances and durations.

Such a theory-relative account of durations is within easy reach of tense primitivism. All we need to do is to adopt a language that can describe the relevant features of the measurement procedures and scientific theories on which our temporal coordinate system is based. In that case, the intrinsic nature of two ersatz times would determine the temporal distance between them. Like the earlier-than relation $\lhd$, temporal distances between ersatz times would be *internal* relations, and they would be as *objective* as such relations can get.

8

TIME AND EXISTENCE

ADDING quantifiers and identity to a system of modal logic is known to be a non-trivial task. W. V. Quine famously argued that there is no coherent account of the interaction of modal operators and quantifiers that does not lead us straight back to the 'metaphysical jungle of Aristotelian essentialism' (1953: 176). These worries were later shown to be mistaken,[57] but this does not change the fact that there are a number of philosophically interesting question that a quantified modal logic needs to settle. Since similar issues arise in the tense case, we need to expand our simple tense logic Z to provide an account of the interaction of tense operators with quantifiers and identity.

8.1 The logic of quantifiers

To set up the discussion, let me begin with a quick survey of quantificational logic. Suppose we are given a formal language with individual constants a, b, c, d, e, variables x, y, z, and predicates F, G, H, I, J, K of various adicities. To ensure an inexhaustible supply of them, numerals may be added as subscripts to constants, variables, and predicates. As logical constants, we use the identity symbol $=$ and the existential quantifier $\exists$ in addition to the sentence connectives $\neg$ and $\rightarrow$ that we have already. As usual, a *term* is either a variable or an individual constant. An atomic formula is then either an n-ary predicate followed by n terms, or the identity sign flanked by two terms. Molecular formulae are obtained by combining atomic formulae with the help of the logical constants. Both $\neg\varphi$ and $\varphi \rightarrow \psi$ are formulae if φ and ψ are, and if φ is a formula and ξ a variable then $\exists\xi\,\varphi$ is a formula. In such a case, the quantifier is said to bind all occurrences of ξ within φ, unless the occurrence is already bound by an earlier quantifier. An occurrence of a variable not bound by any quantifier is called free, and any formula without free variables is a sentence. Universally quantified formulae of the form $\forall\xi\,\varphi$ are abbreviations for $\neg\exists\xi\,\neg\varphi$.

[57] See Marcus (1993) and Parsons (1967; 1969).

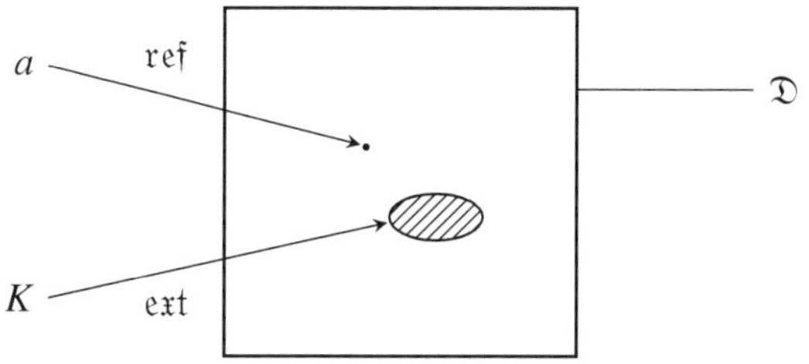

Figure 8.1. Models for quantificational logic

A model for this language is a triple $\mathfrak{m} = \langle \mathfrak{D}, \mathfrak{ref}, \mathfrak{ext} \rangle$ that consists of a set of objects $\mathfrak{D}$ chosen as *domain* and two maps $\mathfrak{ref}$ and $\mathfrak{ext}$. The *referent assignment* $\mathfrak{ref}$ provides every individual constant with an element of the domain $\mathfrak{D}$ as its referent. Some objects in the domain may go unnamed, and some may get more than one name, but no individual constant can fail to name something in the domain. The *extension assignment* $\mathfrak{ext}$ does something similar for predicates. It selects a set of n-tuples of elements of $\mathfrak{D}$ for each n-ary predicate other than identity. Each unary predicate is assigned a set of objects from the domain, each binary predicate a set of ordered pairs of objects, and so on. The same set might serve as the extension of more than one predicate, but not every subset of the domain needs to serve as the extension of a predicate.

Suppose that Φ is an n-ary predicate and $\alpha_1, \ldots, \alpha_n$ a sequence of n not necessarily distinct individual constants. Then the atomic sentence $\Phi\alpha_1 \ldots \alpha_n$ is true in the model $\mathfrak{m}$ if and only if the n-tuple of objects $\langle \mathfrak{ref}(\alpha_1), \ldots, \mathfrak{ref}(\alpha_n) \rangle$ assigned to the constants is an element of the set of n-tuples $\mathfrak{ext}(\Phi)$ assigned to the predicate. Identity is a binary predicate, but its interpretation is not provided by the extension assignment. The interpretation of identity is fixed by the stipulation that an atomic sentence of the form $\alpha_1 = \alpha_2$ is true in a model if and only if $\mathfrak{ref}$ assigns the same referent to the individual constants α_1 and α_2. This definition extends to truth-functional compounds in the obvious way and tells us to what it is for sentences without quantifiers to be true in $\mathfrak{m}$:

$$
\begin{array}{lll}
\mathfrak{m} \models \Phi\alpha_1 \ldots \alpha_n & \text{iff} & \langle \mathfrak{ref}(\alpha_1), \ldots, \mathfrak{ref}(\alpha_n) \rangle \in \mathfrak{ext}(\Phi) \\
\mathfrak{m} \models \alpha_1 = \alpha_2 & \text{iff} & \mathfrak{ref}(\alpha_1) = \mathfrak{ref}(\alpha_2) \\
\mathfrak{m} \models \neg\varphi & \text{iff} & \text{it is not the case that } \mathfrak{m} \models \varphi \\
\mathfrak{m} \models \varphi \rightarrow \psi & \text{iff} & \text{either } \mathfrak{m} \models \neg\varphi \text{ or } \mathfrak{m} \models \psi
\end{array}
$$

In the example depicted in Fig. 8.1, $\mathfrak{m} \models Ka$ is false because the object that $\mathfrak{ref}$ assigns to the individual constant a is not contained in the subset of the domain that $\mathfrak{ext}$ assigns to the monadic predicate K.

The interpretation the individual constants is fixed by the referent assignment $\mathfrak{ref}$. To obtain something similar for the variables, let a *variable assignment* $\mathfrak{var}$ be a map that assigns an element of the domain to each variable. Jointly, $\mathfrak{ref}$ and $\mathfrak{var}$ define the *denotation map* $\mathfrak{den}$, which assigns to each term τ an element of the domain:

$$\mathfrak{den}(\tau) = \begin{cases} \mathfrak{ref}(\tau) & \text{if } \tau \text{ is a constant} \\ \mathfrak{var}(\tau) & \text{if } \tau \text{ is a variable} \end{cases}$$

Relative to a variable assignment $\mathfrak{var}$, an atomic formula $\Phi\tau_1 \ldots \tau_n$ is true in a model just in case the n-tuple of the denotations of the terms $\tau_1, \ldots, \tau_n$ is contained in the set of n-tuples that $\mathfrak{ext}$ assigns to the n-ary predicate Φ. An identity statement $\tau_1 = \tau_2$ is true relative to a variable assignment if and only if the terms τ_1 and τ_2 have the same denotation. This again extends to truth-functional compounds:

$$\begin{array}{lll} \mathfrak{m} \models_{\mathfrak{var}} \Phi\tau_1 \ldots \tau_n & \text{iff} & \langle \mathfrak{den}(\tau_1), \ldots, \mathfrak{den}(\tau_n) \rangle \in \mathfrak{ext}(\Phi) \\ \mathfrak{m} \models_{\mathfrak{var}} \tau_1 = \tau_2 & \text{iff} & \mathfrak{den}(\tau_1) = \mathfrak{den}(\tau_2) \\ \mathfrak{m} \models_{\mathfrak{var}} \neg\varphi & \text{iff} & \text{it is not the case that } \mathfrak{m} \models_{\mathfrak{var}} \varphi \\ \mathfrak{m} \models_{\mathfrak{var}} \varphi \rightarrow \psi & \text{iff} & \text{either } \mathfrak{m} \models_{\mathfrak{var}} \neg\varphi \text{ or } \mathfrak{m} \models_{\mathfrak{var}} \psi \end{array}$$

To account for quantified sentences of the form $\exists\xi\,\varphi$, say that a variable assignment $\mathfrak{var}^*$ is a ξ-*variant* of the variable assignment $\mathfrak{var}$ if and only if $\mathfrak{var}$ and $\mathfrak{var}^*$ agree everywhere, with the possible exception of what they assign to the variable ξ. We can then say:

$$\mathfrak{m} \models_{\mathfrak{var}} \exists\xi\,\varphi \text{ iff } \mathfrak{m} \models_{\mathrm{var}^*} \varphi \text{ for some } \xi\text{-variant } \mathfrak{var}^* \text{ of } \mathfrak{var}.$$

This ensures that quantifiers range over all objects in the domain $\mathfrak{D}$, including those that our denotation map does not assign to any term. A formula is said to be true in a model, $\mathfrak{m} \models \varphi$, if and only if $\mathfrak{m} \models_{\mathfrak{var}} \varphi$ holds for all variable assignments $\mathfrak{var}$. Logical truth and consequence are defined as usual, in terms of truth in a model. For sentences, the relativization of $\models$ to a variable assignment makes no difference. If a sentence is true on one valuation then it is true on all of them. The matter is different for formulae with free variables, which are interpreted as containing an implicit generality claim. For example, the formula $x = x$ makes the same claim as the sentence $\forall x\, x = x$.

A proof theory for this quantificational logic can be obtained by adding the following axioms and the inference rule of universal generalization to the propositional logic comprised of A1–A3 and MP:

Q1 $\forall\xi\,\varphi \rightarrow \varphi[\tau/\xi]$
Q2 $\forall\xi\,(\varphi \rightarrow \psi) \rightarrow (\varphi \rightarrow \forall\xi\,\psi)$ provided ξ is not free in φ
Q3 $\tau = \tau$
Q4 $\tau_1 = \tau_2 \rightarrow (\varphi \rightarrow \varphi[\tau_1/\!/\tau_2])$
UG If $\vdash \varphi$ then $\vdash \forall\xi\,\varphi$

Here $\varphi[\tau/\xi]$ denotes the result of substituting the term τ for *every* occurrence of the variable ξ in the formula φ, and $\varphi[\tau_1/\!/\tau_2]$ the result of substituting *some* (but perhaps all) occurrences of the term τ_2 with the term τ_1. Q4 is a schematic version of the Indiscernibility of Identicals. One can then prove as soundness and completeness theorem that a sentence of quantificational logic is a logical consequence of a set of premises if and only if it is provable from them in the system with axioms A1–A3 and Q1–Q4 and inference rules MP and UG.

8.2 Operators and quantifiers

If we want to add tense operators to such a quantificational logic then we need to specify how to evaluate the quantifier at different times. There are two main views we could adopt. Prior (1967: ch. VIII) advocates a *time-relative* account on which the domain of quantification varies from time to time. At each time, the quantifiers only range over the objects that exist then. I favour the *untensed* quantifier view of Nicholas Rescher (1966: sec. IX) and Nino Cocchiarella (2007: ch. 2), according to which the range of quantification always comprises all objects that exist at some time or other.

In the modal case, the view that the quantifiers have different domains in different possible worlds works reasonably well. The same cannot be said about the time-relative quantifier view. Consider David Lewis' (2004) example 'There were three kings named Charles'. If we tried to regiment this sentence as $\exists x \exists y \exists z\, \mathsf{P}(\ldots)$ then its truth would require the present existence of three people named Charles who were kings in the past, thus incorrectly counting the sentence as false. The opposite combination does not work, either, because $\mathsf{P}\exists x \exists y \exists z(\ldots)$ asserts that there was a past time at which there were three kings named Charles simultaneously, which is also false. To formalize this sentence, we need an untensed quantifier that ranges over all objects that have existed in the past, do exist now, or will exist in the future: $\exists x \exists y \exists z(x \neq y \wedge y \neq z \wedge z \neq x \wedge \mathsf{P}(Kx \wedge Cx) \wedge \mathsf{P}(Ky \wedge Cy) \wedge \mathsf{P}(Kz \wedge Cz))$.[58]

If we adopt untensed quantifiers then we need an independent *existence predicate* E! to render time-relative existence claims. In ordinary quantificational logic, we can define such a predicate by taking E!x to be shorthand for $\exists y\, y = x$. In quantified tense logic, we can only define the existence predicate in this way if $\exists$ is a time-relative quantifier that has different ranges at different times. If we read it as an untensed quantifier then every object that exists at some time or other satisfies $\exists y\, y = x$, but only those that exist at the time in question satisfy E!x.

Nino Cocchiarella (2007: ch. 2) deals with this issue by introducing two quantifiers: one that ranges over presently existing objects, and which could be used to define E!, and a second quantifier that ranges over all objects. A third route is suggested by Timothy Williamson (1998; 2002). Rather than introduce a logically independent existence predicate, Williamson's approach would continue to treat E!x as an abbreviation for $\exists y\, y = x$ and conclude that everything always exists. Objects merely lack spatial and other properties at those times when the philosophically unenlightened say that they do not exist.

Another issue concerns the interpretation of individual constants. If we add Q1–Q4 and UG to our tense logic Z then all individual constants must be *temporally rigid*. Unlike definite descriptions like 'the Chancellor of Germany', whose referent changes over time, our constants would have to denote the same object at all times. Here is a quick proof. The Indiscernibility of Identicals Q4 yields $a = x \rightarrow (\mathsf{H}x = x \rightarrow \mathsf{H}a = x)$. Since $\mathsf{H}x = x$ follows from Q3 by the temporal generalization rule TG, we get $a = x \rightarrow \mathsf{H}a = x$ by propositional logic. A similar result holds for the future-tense operator G, or any arbitrary sequences of G and H.

If we regard individual constants as formal counterparts of proper names in English then this view is quite plausible. The same kind of linguistic evidence that Saul Kripke (1980) uses to support the view that proper names are modally rigid support their temporal rigidity as well. If that is right, though, then not any arbitrary collection of models of the

[58]Untensed quantifiers also allow us to regiment Kamp's sentence 'A child was born that will become ruler of the world' from p. 57 as $\exists x(\mathsf{P}Bx \wedge \mathsf{F}Rx)$, without making use of the two-dimensional 'now' operator N. Meyer (2009a) proves that all occurrences of 'now' and 'then' are eliminable in a tense logic with sufficient quantificational resources. What these two-dimensional operators contribute to tense logic is the ability to rigidify descriptions. The same effect can be achieved with untensed quantifiers.

underlying quantificational logic qualifies as a model of quantified tense logic. To guarantee temporal rigidity, we could only combine models with the same referent assignment.

A related problem is that one can use proper names in English to make true claims about objects at times at which they do not exist. 'Vincent van Gogh is famous' is true now, in spite of the fact that 'Vincent van Gogh' presently lacks a referent. While he existed, van Gogh was not famous at all. Unless we want to adopt the implausible view that what is famous now is not van-Gogh-the-person, but some presently existing van-Gogh-idea, we need to make a distinction between two types of predicates. Some predicates are *existence entailing* in that they can only be truly attributed to objects at times at which they exist. Existence-entailing predicates work like the predicates in ordinary quantificational logic, but our quantified tense logic also needs to make room for predicates that are not existence-entailing, like 'is famous'.[59]

8.3 A quantified tense logic

We need a quantified tense logic that deals with all of the issues raised in the previous section. In the modal case, some authors want to modify the underlying quantificational logic, by giving up the full strength of the universal generalisation rule UG. My proposal is to leave quantificational logic alone, and to adopt a tense analogue of Bernard Linsky and Edward Zalta's *simplest quantified modal logic*. This approach locates all the complications in the model theory and allows us to construct a complete system of quantified tense (or modal) logic by combining standard axioms and rules for sentential operators with standard axioms and rules for quantifiers and identity.[60]

Let me begin by describing the model theory of such a quantified tense logic. A model $\langle \mathfrak{D}, \mathfrak{T}, <, \mathfrak{ref}, \mathfrak{ext}, \mathfrak{p} \rangle$ consists of a set of objects $\mathfrak{D}$ that serves as the range of the untensed quantifiers and a set of objects $\mathfrak{T}$ that serve as times. The set $\mathfrak{T}$ is equipped with a binary relation $<$ and one element $\mathfrak{p}$ of $\mathfrak{T}$ is chosen as the present time of the model. The extension map $\mathfrak{ext}$ now assigns a set of n-tuples of elements of $\mathfrak{D}$ to each n-ary predicate Φ *and* time $\mathfrak{t}$ in $\mathfrak{T}$, thus allowing predicates to have different extensions at different times. The referent assignment $\mathfrak{ref}$ assigns

[59]See Prior (1967: 161), Cocchiarella (1969), Woods (1976), and Chisholm (1990).

[60]Linsky and Zalta (1994). A survey of other systems of quantified modal logic can be found in Hughes and Cresswell (1996) and Garson (2001).

an element of $\mathfrak{D}$ to each individual constant, but is not relativized to times, which ensures the temporal rigidity of individual constants.

The existence predicate also gets interpreted by the extension assignment. For each $t \in \mathfrak{T}$, this yields the set $\mathfrak{D}_t = \mathfrak{ext}(E!, t)$ of all objects that exist at that time. When interpreting other predicates, $\mathfrak{ext}$ specifies for all objects in $\mathfrak{D}$ whether or not they have a given property at a given time t, not only for those that happen to be contained in $\mathfrak{D}_t$. This allows us to attribute properties to objects at times at which they do not exist.

If we define the denotation of a term as before then we can give the following definition of what it is for a formula to be true at a time in a model relative to a variable assignment:

$$
\begin{array}{lll}
t \models_{\mathfrak{var}} \Phi\tau_1 \ldots \tau_n & \text{iff} & \langle \mathfrak{den}(\tau_1), \ldots, \mathfrak{den}(\tau_n) \rangle \in \mathfrak{ext}(\Phi, t) \\
t \models_{\mathfrak{var}} \tau_1 = \tau_2 & \text{iff} & \mathfrak{den}(\tau_1) = \mathfrak{den}(\tau_2) \\
t \models_{\mathfrak{var}} \neg\varphi & \text{iff} & \text{it is not the case that } t \models_{\mathfrak{var}} \varphi \\
t \models_{\mathfrak{var}} \varphi \rightarrow \psi & \text{iff} & \text{either } t \models_{\mathfrak{var}} \neg\varphi \text{ or } t \models_{\mathfrak{var}} \psi \\
t \models_{\mathfrak{var}} \mathsf{F}\varphi & \text{iff} & t' \models_{\mathfrak{var}} \varphi \text{ for some } t' \in \mathfrak{T} \text{ such that } t < t' \\
t \models_{\mathfrak{var}} \mathsf{P}\varphi & \text{iff} & t' \models_{\mathfrak{var}} \varphi \text{ for some } t' \in \mathfrak{T} \text{ such that } t' < t \\
t \models_{\mathfrak{var}} \forall\xi\,\varphi & \text{iff} & t \models_{\text{var}^*} \varphi \text{ for all } \xi\text{-variants } \mathfrak{var}^* \text{ of } \mathfrak{var}
\end{array}
$$

A formula is true at a time if it is true under all variable assignments; it is true in a model if it is true at its present time.

Section 4.3 treated models of propositional tense logic as stacks of models of the untensed propositional logic. This result does not extend to the quantificational case. Suppose we take a model of quantified tense logic and restrict it to a single time: $\langle \mathfrak{D}_t, <, \mathfrak{ref}, \mathfrak{ext}(\ \ , t) \rangle$. Since not every individual constant needs to have a referent within $\mathfrak{D}_t$, this restriction does not yield a model of ordinary quantificational logic, but what is known as an *outer-domain model* for free logic.[61] While any suitably arranged collection of models of propositional logic qualifies as a model of propositional tense logic, an arbitrary collection of even such outer-domain models only constitutes a model of quantified tense logic if their outer domains and referent assignments coincide. The only thing they may disagree on are the extensions of predicates.

This model theory for quantified tense logic might be a little bit complicated, but an axiomatic system for such a logic can be obtained by combining the axioms for tense operators with those of our system of untensed quantificational logic. We start out with A1–A3 and MP to

[61] See, e.g. Bencivenga (2002: sec. 8).

govern truth-functional sentence connectives. To this base system, we add Q1–Q4 and UG for quantifiers and identity and Z1, Z2, and TG for tense operators. We also need two axiom schemata that specify how quantifiers interact with tense operators:

$$\text{BF} \quad \mathsf{P}\exists\xi\,\varphi \leftrightarrow \exists\xi\,\mathsf{P}\varphi \qquad \mathsf{F}\exists\xi\,\varphi \leftrightarrow \exists\xi\,\mathsf{F}\varphi$$

These are tense analogues of the modal principle $\exists\xi\Diamond\varphi \leftrightarrow \Diamond\exists\xi\varphi$ due to Ruth Barcan (1946). On a time-relative reading of the quantifiers, these tensed Barcan formulae falsely assert that everything that did or will exist exists now, but both are logical truths on an untensed reading. If the domain of quantification does not change over time then quantifiers and tense operators commute. This system of quantified tense logic consists of two copies of Linsky and Zalta's simplest quantified modal logic, plus the axiom Z2 to ensure that earlier-than is the converse of later-than. By plagiarizing their soundness and completeness proof, one can then easily show that a sentence is a theorem of this system if and only if it is true in all models of quantified tense logic.

Since our quantified tense logic is an extension of Z, the construction of ersatz time series works in exactly the same way as before, but there is one minor issue that needs to be addressed. Section 6.2 noted that time-series can do all the logical work of models in the proof of the soundness and completeness theorem. To extend this to the quantified case, we need to show that any time series T_p yields a model of quantified tense logic in which the same sentences are true.

Here is a quick sketch how this can be done, by adapting the construction of canonical models for quantified modal logic.[62] We begin by expanding our language with countably many extra individual constants. These constants are used to construct a maximal consistent set of sentences $\overline{p}$ that contains p and satisfies the following condition: at every time of the resulting ersatz time series $T_{\overline{p}}$, every existentially quantified sentence has a witness that is named by some individual constant. That is, whenever $t \in T_{\overline{p}}$ and $t \mid \exists\xi\,\varphi$, there is an individual constant α such that $t \mid \varphi[\alpha/\xi]$. With this in place, say that two individual constants α_1 and α_2 are equivalent if and only if $\overline{p} \mid \alpha_1 = \alpha_2$, and let the equivalence class $[\alpha]$ be the set of all individual constants that are equivalent to α. As the domain D of our model we then take the set of all such equivalence classes of individual constants. The referent assignment maps

[62] See, e.g. Hughes and Cresswell (1996: chs. 14, 17).

each constant to its equivalence class, $\mathrm{ref}(\alpha) = [\alpha]$, and the extension assignment is defined by stipulating that $\langle[\alpha_1], \ldots, [\alpha_n]\rangle \in \mathrm{ext}(t, \Phi)$ if and only if $t \mid \Phi\alpha_1 \ldots \alpha_n$. Here $\alpha_1, \ldots, \alpha_n$ are individual constants, t is an element of T_p, and Φ is an n-ary predicate. The relation $\lhd$ on $T_{\overline{p}}$ is defined as on p. 61. This yields a model $\langle D, T_{\overline{p}}, \lhd, \mathrm{ref}, \mathrm{ext}, \overline{p}\rangle$ of quantified tense logic in which all sentences of the original language, prior to the addition of the extra constants, have the same truth-values as in T_p.

This construction does not yield a unique model. There are many other models of quantified tense logic that yield the same ersatz time series. Some of these models can be obtained by replacing our equivalence classes of individual constants with different abstract objects as the domain. If we are lucky, and if the sentences in p are in fact true, then we might also get an *intended* model in which constants name material objects and in which predicates pick out genuine properties, but logic alone cannot tell us whether there is such a 'concrete' model.

9

PRESENTISM

One of the more controversial questions about time is whether there is anything metaphysically special about the present moment. There are three main views about this issue. *Presentism*[63] claims that only present objects exist; the *growing block*[64] view holds that only the past and present are real, but not the future; and *eternalism* contends that all times are metaphysically on the same footing. Apart from being of evident interest in its own right, this question matters to tense primitivism because it is widely thought that any theory of time that is formulated in terms of primitive tense operators is thereby committed to presentism. This chapter reaches the opposite conclusion: tense primitivists, like everybody else, must endorse eternalism.

9.1 The triviality objection

The main challenge to presentism is thought to originate from modern physics. Eternalists argue that presentism is incompatible with the theory of relativity and is thus false a posteriori.[65] According to the theory of relativity, there are no facts about simultaneity that are independent of our choice of reference frame, and hence no frame-independent facts about what objects presently exist. Since what is real should not depend on an arbitrary choice of reference frame, it follows that what is real cannot coincide with what is present. While the theory of relativity seems openly hostile to the doctrine, even classical mechanics is at best indifferent to presentism; it never even mentions the present and treats all times on the same footing. Whatever else we may say about it, presentism is therefore not a scientific thesis.

[63] See, e.g. Lucretius (1947: bk. I), Augustine of Hippo (1999: ch. 11), Hobbes (1651: pt. I, ch. 3), and Schopenhauer (1859: bk. IV, §54). More recently, the view was defended by Mead (1932), Prior (1972), Bigelow (1996), Markosian (2004), Crisp (2004), Ludlow (2004), Bourne (2006), Merricks (2007), Zimmerman (2008), and others. Sider (1999) is impressed by the ontological benefits of presentism, but rejects it in (2001: ch. 2).

[64] Broad (1923) and Tooley (1997) defend versions of this view.

[65] See Rietdijk (1966), Putnam (1967; 2008), Stein (1968), Weingard (1972), Sklar (1981), Savitt (1998), Saunders (2002), and Balashov and Janssen (2003).

This is an important point, but I also think that such an appeal to physical theory grants the presentist too much, by conceding that there is a substantial thesis that needs to be refuted by sophisticated scientific inquiry. As it is usually formulated, presentism is supposed to be the following ontological thesis about time and existence:

(P) Nothing *exists* that is not present

As it stands, this claim is ambiguous between two different readings of the word 'exists' that occurs in it. One option is to regard it as the present tense of the verb *to exist*:

(P1) Nothing *exists now* that is not present

Though true, this claim is utterly trivial. To exist now and to be present are the very same thing. The quantified tense logic from Chapter 8 would formalize both in terms of the existence predicate, and presentism would turn out to be nothing more than a simple logical truth:

$$\underbrace{\neg\exists x}_{\text{Nothing}} \quad (\underbrace{\mathrm{E}!x}_{\text{exists now}} \quad \wedge \quad \underbrace{\neg\mathrm{E}!x}_{\text{that is not present}})$$

Given that presentists clearly meant to advance a substantial thesis, this is presumably not what they had in mind. But what could they have meant? The apparent problem with (P1) is that it interprets 'exists' in too narrow a sense. One might hope for a more plausible view by using a more inclusive notion of existence. Say that an object exists *temporally* just in case it exists at some time or other. This yields a second reading:

(P2) Nothing *exists temporally* that is not present

Since it does not follow from an object's existing temporally that it exists now, this is a non-trivial thesis. If we simplify matters by ignoring the possibility of branches then (P2) can be formalized as follows:

$$\underbrace{\forall x}_{\text{Everything}} \quad ((\underbrace{\mathsf{P}\mathrm{E}!x \vee \mathrm{E}!x \vee \mathsf{F}\mathrm{E}!x}_{\text{that exists temporally}}) \rightarrow \underbrace{\mathrm{E}!x}_{\text{is present}})$$

This sentence is true in some models of quantified tense logic and false in others. Alas, it is easily seen to be false in the *intended* model, in which the variables range over all actual objects. Consider:

(JC) Julius Caesar crossed the Rubicon

Using obvious abbreviations, this true sentence can be formalized as $\mathsf{P}Rc$. Non-existent people might be admired or reviled, but they cannot cross rivers, which means that R is existence-entailing in the sense of p. 83. Hence $Rc \rightarrow \mathrm{E!}c$ is necessary, and thus always true, which yields $\mathsf{P}Rc \rightarrow \mathsf{P}\mathrm{E!}c$ with the tense axiom Z1. $\mathsf{P}Rc$ thus entails that Caesar did exist, $\mathsf{P}\mathrm{E!}c$, and hence that he does exist temporally, $\mathsf{P}\mathrm{E!}c \vee \mathrm{E!}c \vee \mathsf{F}\mathrm{E!}c$ by disjunction introduction. With (P2), it follows that Caesar exists now, $\mathrm{E!}c$, which is clearly false since he is known to have been assassinated in 44 BC. Hence there is an object, Julius Caesar, that exists temporally without being present, and (P2) is false.

Both (P1) and (P2) take the 'exists' in (P) to encode some temporal information. Presentists might protest that neither reading is an acceptable interpretation of their thesis. With 'exists', they mean to express existence *simpliciter*. But suppose we are given an account of existence simpliciter (whatever it may be), and suppose that a exists in this sense. Then a either exists in the actual world or it exists in some other possible world. And if a exists in the actual world then it either exists temporally, at *some* time, or 'outside' time altogether. No matter what account of existence simpliciter we pick, the following conditional is necessary because its consequent exhausts all the ways in which a could exist:

a exists simpliciter → (a exists temporally
∨ a exists outside time
∨ a exists in some other possible world)

Existence simpliciter can only differ from temporal existence by including some objects outside time, some merely possible ones, or both. But to say that nothing exists outside time that is not present is just a roundabout way of denying that there are objects outside time, and to say that nothing exists in other possible worlds that is not present is at best a way of denying that there are merely possible objects. Both might be nontrivial, but neither is a presentist thesis. The only respects in which existence simpliciter could differ from temporal existence are thus irrelevant for our purposes. Temporal existence is already the most general notion of existence that needs to be considered.

Hence presentism is either trivially true, by claiming that nothing exists now that is not present (P1); obviously false, by denying that anything has existed or will exist that does not exist now (P2); or a combination of the two. On none of these reading does (P) express a substantial truth about time and existence.[66] The only way of preserving the illusion that presentism is both true and non-trivial is by constantly ambiguating between (P1) and (P2). This is a problem for presentism, but not for tense primitivism. There is no reason why an advocate of a modal view of time should either deny (P1) or endorse (P2).

Thomas Crisp (2003: 213) argues that this triviality objection to presentism trades on a *de re–de dicto* confusion. One way of reading the claim that nothing has existed that is not present is as a *de dicto* claim:

$$\neg \mathsf{P}\exists x(\mathrm{E}!x \wedge \neg \mathsf{N}\mathrm{E}!x)$$

Here N is the 'now' operator from p. 57. Crisp admits that this sentence is clearly false because it conflicts with true claims like (JC), but he argues that the situation is different for the following *de re* claim:

$$\neg \exists x \mathsf{P}(\mathrm{E}!x \wedge \neg \mathsf{N}\mathrm{E}!x)$$

I do not think this is correct. If we give the quantifiers a time-relative reading then Crisp's *de re* claim is trivially true; it would assert that nothing exists now that does not exist now. Yet if we use untensed quantifiers then we can rewrite the claim as $\forall x \neg(\mathsf{P}\mathrm{E}!x \wedge \neg \mathrm{E}!x)$ and instantiate with the individual constant c to get the obviously false conclusion $\neg(\mathsf{P}\mathrm{E}!c \wedge \neg \mathrm{E}!c)$, which claims that Caesar either never existed or that he does exist now. Moreover, if we use untensed quantifiers then the *de re* and the *de dicto* reading are logically equivalent by the tensed Barcan Formula BF from p. 85 and cannot have different truth values. If the *de dicto* claim is false then the same must hold for the *de re* claim.

Others might object that the real issue in assessing (P2) is whether names like 'Julius Caesar' have a referent. By using the existential generalization from (JC) to 'There is something that exists temporally', I seem to assume that the question has already been settled in my favour. This inference is licensed by the quantified tense logic from Chapter 8, but suppose we permit existential generalizations only for names that presently have a referent. By imposing such a present-referent condition we would in effect adopt a time-relative reading of the quantifiers, which means that the notion of existence employed is really *exists now*,

[66] Versions of this triviality objection can be found in Lombard (1999), Callender (1998), Meyer (2005b), and Savitt (2006). See also Dorato (2006) and the exchange between Crisp and Ludlow in Zimmerman (2004).

and we are back to the trivial thesis (P1) that nothing exists now that is not present. The problem for presentism thus arises independently of our views about the reference. Whatever we say about the semantic value of proper names used at times at which their referents do not exist, any account that ends up rejecting all historical claims like (JC) as either false or meaningless is *eo ipso* unacceptable as a theory of names.

Advocates of the growing block view will point out that this triviality objection only provides an argument against presentism, and that their own account escapes unharmed. Presentists cannot avoid the difficulties with (JC) by denying all contingent truths about the past; nothing more would need to be said about such a view than to repeat that it does *deny*, on purely philosophical grounds, that Caesar crossed the Rubicon.[67] But growing blockers are happy to reject all contingent truths about the *future*, for that is precisely their view about time.

While the growing block view might avoid the triviality objection, it runs into an equally serious difficulty that is entirely of its own making. As Craig Bourne puts the problem, the growing block view leaves us without a plausible account of how we know that it is now now.[68] Unlike competing proposals, the growing block view admits two different senses of 'now'. There is the ordinary indexical now_i, tokens of which refer to the time at which they occur, and the non-indexical now_b, which refers to the time at which the edge of the growing block is located. Bourne's question is how we know that now_i is now_b. Whatever reasons we might give for thinking that our now_i is now_b, the very same reasons were had by people in the past. But past people are located within the growing block and are thus mistaken if they believe that their now_i is now_b. Since their epistemic situation is no different from ours, it follows that we can have no good reasons for thinking that our now_i is now_b.

The relation between the two nows is not a problem for other views about time and existence. Eternalists believe that past, present, and future are equally real. For them, 'now_b' is an empty name because there is no 'edge of reality' to be singled out. Presentists would claim that 'now_i' and 'now_b' are guaranteed to co-refer because neither past nor future exist. For growing blockers, on the other hand, the question is

[67]To paraphrase Lewis (1991: 59), how would you like the job of telling historians that they must change their ways, and abjure countless errors, now that *philosophy* has discovered that Caesar did not exist?

[68]Bourne (2002). Similar issues are raised in Braddon-Mitchell (2004) and Merricks (2006). See also Forrest (2004), Button (2006; 2007), and Tallant (2007; 2011).

both non-trivial and embarrassing. Unless there is a way of identifying now_b, it is hard to see what explanatory work the growing block view could possibly perform that would speak in favour of its adoption.

More seriously, the subject matter of a theory of presentness gets fixed by the indexical sense of now, and it is not a theory of *presentness* unless it tells us what is metaphysically special about *that* time. We might adopt a birds-eye view on the time series, but this does not change the fact that we find ourselves within time. When we ask whether the present is metaphysically privileged, we ask whether there is something special about *our* parochial perspective on the time series. Growing blockers are unable to provide such an account.

9.2 Reconstructive presentism

Some presentists try to rescue (P2) by offering claims about present objects as substitutes for problematic claims about past objects like (JC). Such proposals might deliver theories with the right sort of ontological commitment, but they invariably end up endorsing implausible linguistic theses about the meanings of past tense claims and are unacceptable for that reason alone. Let me illustrate this point with two examples.

Trace presentists argue that past tense claims are about the presently existing causal traces of past events, and that future tense claims are about the present seeds of future events, but without admitting the existence of either type of non-present event.[69] When asserting (JC), we would thus be talking about the present traces of the Rubicon crossing. This would have been the case even right after the event, when Caesar's horse was still wet. A legionnaire, pointing at Caesar and the Rubicon, who says 'This man crossed that river' would not be talking about Caesar, either. Moreover, the causal traces of the event changed quite dramatically throughout history. The traces of the Rubicon crossing first consisted in the brain states of the eye witnesses, then in the brain states of people who merely heard about it, then in some stone carving, then in a long sequence of successive stained parchments, then in a collection of books, then in the electronic state of some computer chip, and so on. If we are to take trace presentism seriously then we would have to say that the subject matter of (JC) constantly shifted, along with the numerous changes in record keeping. That seems patently false; the sentence is surely about Caesar and the Rubicon.

[69]This is the *anti-realism about the past* discussed, and rejected, by Dummett (2003).

Another proposal in this vein is John Bigelow's *Lucretian presentism*. Let *n* be the *now-stuff*, the mereological sum-total of everything that exists at the present moment. What makes (JC) true, Bigelow suggests, is that *n* has the property of being-such-that-Caesar-crossed-the-Rubicon. While his opponents postulate the existence of past and future objects, Bigelow 'trades off this ontological expenditure on individuals against a lavish outlay of properties of present individuals.'[70] Unlike trace presentism, this view does not spell out (JC) in terms of successively different properties of different objects. The property at stake is always being-such-that-Caesar-crossed-the-Rubicon, but the object to which it is being attributed does change quite significantly. Even when the world no longer bears any ordinary traces of the crossing, and when all objects that participated in the event have long gone out of existence, it would still have the property of being-such-that-Caesar-crossed-the-Rubicon. As an account of the meanings of past tense statements, this seems to be even less plausible than trace presentism.

Linguistic intuitions about aboutness are to some extent negotiable. A familiar example are pre-theoretic views about the subject matter of *de re* modal claims. While Saul Kripke (1980) argues that 'Hubert Humphrey might have won the election' is about the actual Humphrey, David Lewis (1986c: ch. 4) thinks that it is about one of his *counterparts* in another possible world. Lewis' proposal might be at variance with our pre-theoretic intuitions, but he argues that its systematic benefits justify a revision of our views about the subject matter of the Humphrey sentence. Taking our linguistic intuitions too seriously, he claims, would lead to serious philosophical difficulties elsewhere.

What is different in our case is that there are no philosophical problems that would be solved by presentism, rather than be created by it. Scientific theory is at best indifferent to presentism. All physical theories, including classical mechanics, treat all times and the objects that exist at them on an equal footing. They never mention properties of presentness, and relativistic theories of motion even seem to rule out their existence. There is no theoretical pressure in favour of a revision of our linguistic intuitions about the meanings of past tense claims.

To be fair, most presentists do not claim that their view serves any such theoretical needs. On the contrary, they think that presentism is

[70] Bigelow (1996: 47); see also Lucretius (1947: bk. I), Keller (2004), Crisp (2007), Zimmerman (2008: 218), and Sanson and Caplan (2010).

part of our commonsensical folk-theory of time, and thus part of the data that a successful metaphysical theory of time needs to account for:

> Nothing exists which is not present. I say that this was believed by everyone, both the philosopher and the folk, until at least the nineteenth century; it is written into the grammar of every natural language; and it is still assumed in everyday life even by philosophers who officially deny it. (Bigelow 1996: 35)

If that is right, though, then we merely get a low-level clash between our intuitions about presentism and our intuitions about the subject matter of (JC), and the triviality objection itself tells us how to resolve this conflict. Given that the reading (P2) of the presentist thesis conflicts with our pre-theoretic views about the meanings of past tense claims, the obvious remedy is to offer the trivially true (P1) as an account of why so many people are initially attracted by the present claim that nothing exists that is not present. Since there is no reason to revise our views about the subject matter of (JC) in order to accommodate (P2), this means that there would be theoretical pressure *not* to take presentism seriously.

The conclusion reached in the previous section can thus stand. Presentism is either trivially true, obviously false, or a combination of the two. There is no reading on which it expresses a substantial truth about time and existence. To show this, we did not need to appeal to any sophisticated physics. All we needed was evidence that there was or will be some object that does not presently exist. Even if Newton had been right about the laws of physics and Einstein wrong, presentism would therefore still have run into the very same problems.

9.3 Presentism and actualism

Many philosophers, including some who reject presentism, worry that the triviality objection would prove too much. Since presentism is the temporal analogue of the modal thesis of *actualism*, they reason, accepting the objection would force us to adopt a similarly dismissive stance about actualism. Since actualism is a non-trivial thesis, they conclude that something must have gone wrong with the triviality objection.[71]

I agree that a mistake is being made, but I think it lies with the assumption that actualism *has* a temporal analogue, not with the triviality objection. Consider the modal counterpart of the presentist thesis (P):

(A) Nothing *exists* that is not actual.

[71] This argument is suggested by Sider (1999: sec. 1), Zimmerman (1998: 211), Hestevold and Carter (2002), and Crisp (2007).

Like (P), this claim is ambiguous between two readings of the 'exists' that occurs in it. If we interpret it as expressing actual existence then we get a trivially true thesis:

(A1) Nothing *actually exists* that is not actual.

On the other hand, if we interpret the 'exists' as expressing possible existence, which is the modal counterpart of temporal existence, then we get a non-trivial thesis that is incompatible with the truism that there could have been objects that do not actually exist:

(A2) Nothing *possibly exists* that is not actual.

It follows that (A) is either trivially true or obviously false.

How do actualists deal with this triviality objection? The short answer is that they *don't*. What is called 'actualism' in the philosophical literature is neither (A) nor any other claim of quantified modal logic. Actualism is a version of the possible-worlds analysis of the modal operators, and thus a claim in the *meta*theory of modal logic. It is a thesis about modal operators, not about the scope of quantifiers.

Amongst all the models of modal logic, there is one that is special, the *intended* model $\langle \Omega, \mathrm{R}, \alpha \rangle$ of modal logic. While we usually speak of what is true in a model, this qualification can be dropped for the intended model. If a statement φ is true at the actual world α of the intended model, $\alpha \models \varphi$, then it is true simpliciter. Applied to sentences with a leading modal operator, the definition of $\models$ from p. 41 yields:

$$\Diamond\varphi \text{ iff } \omega \models \varphi \text{ for some } \omega \in \Omega \text{ such that } \mathrm{R}\alpha\omega$$

If we take S5 as our theory of modality then what is possible does not depend on what is actual. Every possible world is accessible from every other one, and the requirement that ω be accessible from α can be dropped. If we also reserve the term 'possible world' for the elements of the intended model then we can put this more succinctly as:

PWA $\Diamond\varphi$ iff φ is true in some possible world

The *possible-worlds analysis* claims that PWA permits a reductive elimination of modal operators in favour of quantification over possible worlds. The rival view of *modal primitivism* starts at the other end. It

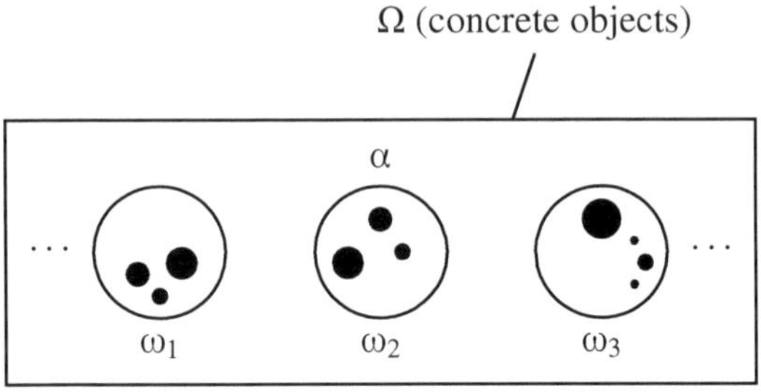

Figure 9.1. Modal realism

takes ◇ as conceptually primitive and tries to spell out all modal notion in terms of this operator, including the notion of a possible world.

What precisely the possible-worlds analysis amounts to depends on what we say about the worlds in the intended model of modal logic. There are two main views to chose from. *Modal realists* like David Lewis (1986c) claim that the intended model consists of a vast number of equally concrete worlds that are spatio-temporally separated from one another, like raisins in a pudding. The actual world α consists of whatever is spatiotemporally connected to us; it is the world that we happen to live in, and other possible worlds are just as concrete.

Actualists deny the existence of the concrete possibilia that make up the possible worlds of the modal realist. Instead, they argue that the possible worlds in the intended model are actually existing, abstract ways the world might have been. One version of this view is the linguistic ersatzism discussed in Section 6.2, which takes possible worlds to be sets of sentences. There are a number of other actualist proposals, but they all claim that the actual world α must be distinguished from *actuality*, which is the mereological sum-total A of all objects that actually exist. Actuality contains the actual world α along with all the other abstract worlds in Ω, but it also contains concrete objects: sticks, stones, etc. The contingent parts of A determine which element of Ω correctly describes the way things actually are and thus counts as the actual world α.

Consider the claim that there might have been pink elephants. According to the modal realist, this claim is true because some other possible world contains concrete, flesh-and-blood pink elephants. For the actualist, it means that there is an abstract way the world might have been according to which there are pink elephants. In spite of their different views about the intended model of modal logic, actualists and

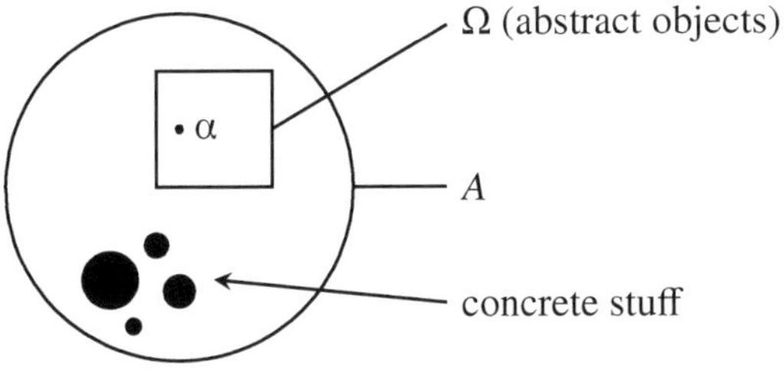

Figure 9.2. Actualism

modal realists agree that (A1) is trivially true and (A2) obviously false. They also agree that there is no intrinsic difference between the actual world and merely possible ones. For the modal realist, other possible worlds are aggregates of concrete objects, just like ours, and all of the actualist's possible worlds are equally abstract. Which of these abstract worlds counts as actual depends on which of them gives the correct description of the contingent parts of actuality. Being actual would be an *extrinsic* feature of a possible world, and not an intrinsic property that would distinguish the actual world from the merely possible ones.

One might suggest that presentists adopt a similar strategy. Rather than try to rescue the trivial thesis (P), they should advocate an elimination of tense operators in favour of presently existing, abstract objects. Suppose we read $\Diamond$ as the tense operator 'sometimes' (p. 43). Let a *possible present* be an abstract way the present might have been. Then the tense analogue of the possible-worlds analysis PWA would be a *possible present analysis* of 'sometimes':

PPA $\Diamond\varphi$ iff φ is true in some possible present

Whatever one might think of PWA, this proposal is clearly untenable. Since the vast majority of possible presents *never* happen, PPA wrongly counts many false claims of the form $\Diamond\varphi$ as true. One could instead say that $\Diamond\varphi$ is true just in case φ is true according to some possible present that is a *time*, but that just brings up the question of what makes it the case that a possible present *is* a time. The obvious answer is that a possible present is a time if and only if everything that is true according to it is sometimes true, but this appeals to the very 'sometimes' operator that PPA tries to eliminate, rendering the proposed analysis circular.

There is no tense analogue of the possible-worlds analysis, and thus no temporal analogue of actualism, either.[72]

In light of the fact that S5 can serve as both a modal logic and a simple tense logic, these differences might be a little bit surprising, but they are easily explained. While S5-as-a-modal-logic and S5-as-a-tense-logic have the same *theorems*, what is at issue in an analysis is the potentially much larger class of all *truths* involving $\Diamond$. The possible-worlds analysis succeeds because there are no *contingent* truths of form $\Box\varphi$ or $\Diamond\varphi$. According to axioms M3 and M4, such claims are only true if they are themselves necessary, and what possible worlds there are does not depend on which world is actual. But if one reads $\Diamond$ as 'sometimes' then the truths of the form $\Diamond\varphi$ outnumber the theorems. Take the S5 axiom M4, $\Diamond\varphi \rightarrow \Box\Diamond\varphi$. Read as an axiom of tense logic, it only demands that any claim that is sometimes true is always sometimes true, not that it is necessarily sometimes true. Since most claims of type $\Diamond\varphi$ are contingent, different possible presents count as times in different possible worlds. The only way of picking out these times amongst all the possible presents is by using the very tense operators that a times-analysis would try to eliminate in favour of quantification over times. What is remarkable is the success of actualism, not the failure of presentism. The availability of the actualist position is due to the fact that it is a theory of possibility and does not reflect a general feature of theories that are cast in terms of intensional operators.

The view of times developed in Chapter 6 already incorporates this lesson. It treats times as presently existing, abstract sets of sentences, but the construction only succeeds because the sentences that make up these times contain the very tense operators that PPA wants to eliminate in favour of quantification over times. Every maximal consistent set of sentences of tense logic contains a maximal consistent set of sentences of the underlying untensed language as a proper subset. What matters for the construction of the time series, though, are the sentences containing tense operators. It is the information contained in these sentences that determines the earlier-than relation $\lhd$, and thus allows us to construct the time series from a given choice of possible present.

The present plays such a prominent role in systems of tense logic because it is *semantically* special. It is the only time that we can refer to by using an indexical ('now') and that we can describe by simply uttering indicative sentences. To refer to other times, we need to use definite de-

[72]Meyer (2006) argues this point in more detail.

scriptions, by employing calendar sentences like the ones discussed on p. 63. In the modal case, Robert Stalnaker (1976) notes that the reasons for accepting or rejecting an indexical account of actuality are independent of the choice between actualism and modal realism. Something similar holds in the tense case, where the fact that the present is semantically special does not entail that it is metaphysically privileged. In this respect, time is quite similar to space. Just as there is no principled difference between *here* and other places, *now* is a time like all others.[73]

[73]This result will become important in Chapter 12, when we turn to the task of reconciling tense primitivism with the theory of relativity.

10

THE FLOW OF TIME

MANY philosophers think that a metaphysics of time must provide an account of the dynamic nature of temporal passage.[74] I disagree. The only coherent sense that can be made of the flowing time metaphor is almost trivial, and does not impose any significant restriction on our account of the nature of time. Just as the present time is no more special than the current location, times are really no more 'dynamic' than places or possible worlds.

10.1 A trivial theory

George Schlesinger claims that 'there is hardly any experience that seems more persistently, or immediately given to us than the relentless flow of time' (1991: 427). I am happy to grant that the passage of time often feels like a river that is inexorably carrying us along, but I do not think that this impression needs to be taken very seriously, or that it would reveal anything remarkable about the nature of time if we did.

What is being perceived are always the *changes* of things in time, not the passage of time itself. When Robin Le Poidevin claims that 'We just *see* time passing in front of us', he immediately equates such passage with 'the movement of a second hand around a clock, or the falling of sand through an hourglass, or indeed any motion or change at all' (2007: 76). If that is all there is to the perception of passage then we do not need a separate account of flow to explain the observational data. Any theory of time that solves the Problem of Change can explain the perception of temporal passage in terms of observed changes.

The same applies to what Oreste Fiocco calls the 'phenomenological novelty of each moment, that is, the continuous sense that one has not encountered *this* state of the world' (2007: 4). If the impression of

[74] See, e.g. McTaggart (1927) and Broad (1923: ch. II). Gale (1968; 1969) initially advocated a similar position but later changed his mind about this. See his (1996) and Garson (1969). Levison (1987) considers the compatibility of a flowing time view and presentism. Also in this case, there is a debate about whether a flowing time is compatible with the theory of relativity; see Stein (1968; 1991) and McCall (1976).

novelty is not mistaken, as it might well be, then something must be true at the present moment that was not true before and these changes alone would then suffice to explain its novelty. The present looks different from earlier times because it *is* different.

There are also epistemic differences between past, present, and future. Suppose we simplify matters by disregarding the time it takes for signals to travel from perceived objects to our sense organs. Then only what presently exists is accessible to perception, our knowledge about the past must rely on memory, and our knowledge about the future on prediction.[75] But these differences only characterize our present epistemic situation and do nothing to distinguish it from other times. At any time, we can only perceive what exists then, our knowledge of earlier times must rely on our memories, and our knowledge of later times on predictions that are possible at that time. If every time is epistemically special then none of them is and there is nothing that a theory of time, rather than a theory of knowledge, needs to account for. As new observations are made and old memories fade, our epistemic situation is in a constant flux, which is just another way of saying that our epistemic situation is constantly changing.

Having said this, there is something peculiar about *how* we perceive changes. Instead of a succession of states with different characteristics, we sometimes perceive change itself. There is a phenomenal difference between seeing the hand of a clock first at one position and then at another, by looking twice, and *seeing the hand move*, by looking only once. William James notes that 'A succession of feelings, in and of itself, is not a feeling of succession. And since, to our successive feelings, a feeling of their own succession is added, that must be treated as an additional fact requiring its own special elucidation' (1890: 628–9). Instead of a sequence of momentary stages, James argues that the content of our perception in such cases must be characterized in terms of what he calls the *specious present*. It is a nice question for a theory of perception whether the specious present is best accounted for in terms of our consciousness being somehow extended in time, or in terms of a retention mechanism by means of which our brains synthesize sensory information received over a short period of time into a momentary impression of change. No matter how we answer this question, though, we

[75] 'The *Present* onely has a being in Nature; things *Past* have a being in the Memory onely, but things *to come* have no being at all, the Future being but a fiction of the mind' (Hobbes 1651: pt. I, ch. 3).

always end up with an account of how beings like us perceive change, and not with an account of the perception of temporal passage itself.[76]

All the observational evidence can thus be accounted for by a *trivial theory of passage* that identifies the flow of time with the occurrence of change.[77] To say that time flows during a time interval would be to say nothing more (but also nothing less) than that change occurs throughout that interval. At each time, something is true then that is false at other times. Such a trivial theory of passage does not encroach on the domain of physics by trying to predict or explain changes, nor does it try to explain why there is more than one time to begin with. Because it regards passage as a feature of what is happening within time, rather than an essential aspect of time itself, such a theory does not really tell us anything about the metaphysics of time.

It is often objected that taking the flow of time seriously would require an account of the *rate* of flow, which is said to lead to the postulation of an infinite sequence of different time series:

> If the claim that time always flows had a literal sense, we ought to be entitled to ask how fast time is flowing. And if so, there would have to be a supertime for measuring the rate of flow of ordinary time. With regard to that super-time, a similar question would immediately arise, viz., how fast it was flowing; and this would imply still another time for its answer. And so on without end.
>
> (Black 1959: 56)

Not everybody thinks that this is a serious problem. Some authors are happy to say that time flows at a rate of one hour per hour, in which case the infinite regress of time series never gets started; others believe that the passage of time only needs a second time series, and that the regress can be brought to a halt at that stage.[78] The trivial theory gives a different answer. If the flow of time consists in the occurrence of changes then the only sense to be made of the rate of passage is in terms of the speed of particular changes. It would be nonsense to ask how fast time passes in general, rather than to inquire about the speed of a particular process. Yet the rate at which a specific change unfolds is just given

[76] See Dainton (2008; 2011), Hoerl (2009), Paul (2010), and Prosser (2010).

[77] This view is similar to the *modest conception of passage* defended by Savitt (2001; 2005). In (2002), he argues that this view is the same as the one Broad (1933: ch. 35) and Williams (1951) arrive at, even though the former regards himself as an advocate of passage and the latter as an opponent.

[78] See Broad (1933: ch. 35), Smart (1949), Webb (1960), Markosian (1993), Price (1996: 13), Maudlin (2002), Olson (2009), Phillips (2009), and especially the long exchange between Schlesinger (1980; 1985; 1991; 1994) and Oaklander (1983; 1984; 1985).

by the temporal distance between the initial state and the changed one. There is no need to postulate any additional time series.

The trivial theory of passage does have one peculiar feature. Take any theory of time that admits the possibility of *freezes*, such as temporal substantivalism or tense primitivism (see the note on p. 68). A freeze is an extended stretch of time during which nothing undergoes change. If we identify the passage of time with the occurrence of change then we would have to say that, from the beginning to the end of such a freeze, time *elapses*, because there are time points between them, but does not *flow*, because nothing changes during the freeze. That might be an odd thing to say, but I think it counts against the view that time flows 'on its own', rather than against the trivial theory. The point of the trivial theory is to pose a challenge to those who believe that there is something metaphysically interesting about the passage of time. The challenge is to explain what the flow of time consists in, if it is not mere change, and to spell out what reason there are for believing that there is such flow, given that the trivial theory already accounts for the observational data.

10.2 The moving NOW

Since change consists of nothing more than an object's possessing a property at one time that it lacks at another, one might complain that the trivial theory is too static, and that it does not do enough to account for the dynamic nature of the flow of time. Steven Savitt, who defends a version of the trivial theory of passage, aptly points out that such concerns are based on a confusion:

> Where in this picture, it may be asked, is passage? My response to this critical question is that change is not in this picture but what it is a picture of. One who asks it is confusing a static representation with a representation of stasis ... We do not need to have an animated picture to have a picture of animation.
>
> (Savitt 2002: 162–3)

Nevertheless, it is an interesting question whether there are coherent accounts of passage that go beyond the trivial view. One proposal that has garnered much attention in the literature is the view that the flow of time consists of the present moving across the time series:

> We are naturally tempted to regard the history of the world as existing eternally in a certain order of events. Along this, and in a fixed direction, we imagine the characteristic of presentness as moving, somewhat like the spot of light from a policeman's bulls-eye traversing the fronts of the houses in a street. What is

illuminated is the present, what has been illuminated is the past, and what has not yet been illuminated is the future. (Broad 1923: 59)

While C. D. Broad presents the series of events as fixed, with the spotlight of the present moving across it, George Schlesinger imagines that all events pass through a stationary NOW in order:

The NOW is something that moves relative to the series of points that constitute time. Temporal points from the future, together with the events that occur at those points, keep approaching the NOW and after momentarily coinciding with it they recede further and further into the past. (Schlesinger 1980: 23)

It does not matter which of these two pictures we adopt; the important point is that there is supposed to be relative movement between the time series and the spotlight or the NOW. A standard criticism of this proposal is the rate-of-passage objection mentioned above, but I think the moving NOW view encounters a far more elementary difficulty.

Suppose we try to spell out the proposal in the context of tense primitivism. (Since parallel issues arise for other accounts of time, nothing of importance depends on this assumption.) Let π be a sentence that is true at a time, $t \mid \pi$, if and only if the moving NOW has arrived at t. Tense primitivists who want to endorse the moving NOW view must admit the existence of such a sentence π, either in the language in which they formulated their view of time or in a suitable extension. Otherwise, their theory of the nature of time would be unacceptably incomplete by failing to account for the passage of time. Next, suppose that the present time is t. Then it is true at t that NOW has arrived, $t \mid \pi$. Since only one time is present, this means that NOW is *not* at any other time, $\neg\exists t'(t' \neq t \wedge t' \mid \pi)$. But if the NOW is indeed moving then it does not stop at t. It will be located at some other time, $t \mid \mathsf{F}\pi$, and thus $\exists t'(t < t' \wedge t' \mid \pi)$ by definition of the future tense operator. So NOW is *also* located at some time t', which contradicts the assumption that it is only located t. The moving NOW view is thus inconsistent.

One might propose using a different sentence π_t for each time t to report that NOW has arrived then. Since the NOW arrives at t at no time other than t, these sentences would satisfy $t' \mid \pi_t \leftrightarrow t = t'$. This is a coherent account, but it is not a theory of passage; it merely claims that different things are true at different times, not that the same thing, namely the presence of the NOW, is exclusively true at successive times.

Instead of using the true-at operator $\mid$ and the earlier-than relation $<$ on times, the *reductio* of the moving NOW view can also be put in terms of primitive tense operators, by using the convention that untensed sentences make claims about the present moment:

π	The present is NOW
$\pi \to \neg \mathsf{F}\pi$	Only one time is NOW
$\mathsf{F}\pi$	NOW moves into the future
$\bot$	Contradiction

Similar problems arise for the claim that NOW was located at another time, $t \mid \mathsf{P}\pi$, which entails $\exists t'(t' < t \wedge t' \mid \pi)$. The NOW's having arrived at a time cannot be something that is true at that time only, but was true at some earlier time, or will be true at some later time.

This kind of problem is familiar from the discussion of theories of actuality. Robert Merrihew Adams considers a proposal that aims to spell out actuality in terms of a primitive property of the one possible world that is actual. This simple property theory of actuality runs into the same difficulties as the moving NOW view:

> Presumably the nonactual worlds could have been actual and are possibly actual. Each possible world is actual in some possible world—namely, in itself. How, then, does the actual world differ from the other possible worlds in relation to the primitive property of actuality? It has the property of actuality, of course, and not just possibly. To have a property *actually* is presumably to have it in the actual world. So the actual world has the property of actuality in the actual world. But that only tells us that the actual world is actual *in itself*. And every possible world is actual *in itself*. So how is the actual world different from the other possible worlds? What is the difference between the actual actual and the possibly actual? Thus, the problem of distinguishing between the actual and the merely possible reappears with respect to the very property of actuality by which it was supposed to be solved. The problem could be solved by a simple property theory only if we were prepared to deny that the nonactual possible worlds are possibly actual. But that denial entails that there is no such thing as contingent actuality. We would have to conclude that the actual world ... is the only possible world that could have been actual. (Adams 1974: 221–2)

On pain of contradiction, being actual cannot be something that is true of the actual world alone, but is possibly true of all other worlds.

In light of these difficulties, Adams abandons the simple property theory and endorses actualism instead. Since we saw in Section 9.3 that actualism has no temporal analogue, this move is not available in the temporal case. But suppose, *per impossible*, that there were a coherent, non-trivial version of presentism. One could then give a presentist account of passage by treating times as presently existing abstract objects.

All of these times represent themselves as being present, but only one of them is right about its own presentness, namely the one that successfully characterizes everything that exists, which presentists believe to include only present objects. That one time is NOW, but as objects change their properties and come into and go out of existence, successively different times satisfy this description. Unfortunately, such an account of passage is no more coherent than the presentism on which it is based.

10.3 Pure becoming

A different attempt at a non-trivial theory of passage is what Richard Taylor (1960; 1974) calls *pure becoming*. As we know from experience, aging usually involves change: wine turns into vinegar, roses wither and slowly turn to dust. These are cases where change merely accompanies aging; according to the doctrine of pure becoming, aging itself is a kind of change. The transition of an object's existing first at one time and then at another is said to amount to a genuine alteration in the object. Tomorrow, all objects will be one day older than they are today, even if they do not change their mass, shape, or any other ordinary property. Pure becoming rules out the possibility of freezes. If time cannot elapse unless all the objects in time experience pure becoming then there can be no extended stretches of time during which no changes occur. Freezes are thus impossible and time cannot elapse without flowing. The passage of time, in the form of pure becoming, would be a precondition for the occurrence of ordinary changes.

One must resist the temptation to think that pure becoming is just the temporal analogue of motion through space. In the case of space, there is a genuine difference between an object's being located first at one place and then at another. For example, to get from point p_1 on the spatial manifold $\mathcal{M}$ to point p_2, an object has to cease bearing the spatial location relation L to p_1 and to start bearing it to p_2. However, in persisting from time t_1 to time t_2 on the temporal manifold $\mathcal{T}$, an object does not cease to bear the exists-at relation E to t_1 and then begin to bear it to t_2. If an object exists at a time then it always exists at that time.

According to relationists about space, motion requires that an object's spatial parts change the spatial relations they bear to the spatial parts of other objects. This is not true for temporal-parts relationism, where a persisting object does not have to shed the part a-at-t_1 before it acquires a-at-t_2. Nor do these parts need to change their temporal relations to each other. For the object to persist, they just need to *be* these temporal parts. On neither of the two spatial views of time does aging mark the progress of an object's journey through time. If anything, persistence would be more similar to the way a stationary object is extended through space, by possessing different parts at different places.

Similar remarks apply to tense primitivism. For a to persist from time t_1 to t_2, the same sentence E!a needs to be true at both times, but there is no property that the object possesses at one time and lacks at the other. Motion is a change with regard to location, but the only change an object can undergo with regard to existence is to be created or destroyed. That neither makes persistence similar to motion nor does it provide a non-trivial way of understanding the flowing time metaphor.

To obtain pure becoming, we need an entirely new class of properties in addition to ordinary properties, such as mass, spin, or charge. Here is one way of obtaining such properties. Say that an object possesses the *anniversary property* $P_{K,n}$ at time t if and only if n time units before t the object first acquired the ordinary property K. If there are such anniversary properties then at least some objects undergo pure becoming. Suppose an object comes into existence at some time and acquires the property K. At any future time, it will then possess an anniversary property that it did not possess before and will never possess again. Only n units of time after its creation does the object have the property $P_{K,n}$. Even if it keeps all of its ordinary properties constant, the object would thus constantly change its anniversary properties.

However, this only gives us genuine change if anniversary properties count as genuine properties, and they are easily seen to lack all of the features that are commonly thought to distinguish bona fide properties from arbitrary ways of grouping objects. For example, genuine properties are supposed to make for genuine *similarity* between the objects that share them. Anniversary properties can fail to do so. We can have two objects that *were* similar n time units ago, when they were both K, but then parted company and are now completely unlike one another, even though they both possess the anniversary $P_{K,n}$.

Anniversary properties also make no difference to the causal powers of the objects that possess them, as demanded by Sydney Shoemaker's (1980) account of properties. Age itself does not have any causal power; that role is played by whatever ordinary properties an object acquired during its aging. Moreover, given that anniversary properties are causally inert, they cannot be perceived, either, which means that the change of anniversary properties also cannot explain our alleged perception of the passage of time.

We could relax our standards for propertyhood and introduce anniversary properties by brute stipulation, but that would not help the case for pure becoming. It was clear from the outset that it is possible for every object in the universe to undergo change at every moment of its existence. One can easily imagine a scenario where every object is equipped with some sort of internal clock that makes it undergo intrinsic change throughout its lifetime. Yet this does not show what needs to be shown, which is that change is *essential* to persistence. None of the accounts of time and change discussed so far is committed to the existence of anniversary properties.

Another way of obtaining pure becoming is by postulating what one might call *tensed properties*. Suppose that all the primitive predicates of the language of tense logic pick out ordinary properties like mass or charge. If these were the only properties there are then an object could persist without changing. But suppose we adopt a more generous account of properties and claim that any formula $\varphi(x)$ of the language of tense logic defines a property. If we assume, for sake of simplicity, that necessarily coextensive properties are identical then this amounts to adopting the following tensed property comprehension schema:

$$\exists X\, \Box \forall x\, (Xx \leftrightarrow \varphi(x))$$

Here X is a property variable and $\Box$ the necessity operator. For a quick illustration of how this theory of properties leads to pure becoming consider the special case where time is discrete, and suppose that we are now at the time t when object a first acquires the property K. Then a has always lacked K in the past, which means that it now has a property defined by the formula $\mathsf{H}\neg Kx$. At the next instant, a will have lost that property and acquired the one defined by $\mathsf{P}Kx$ instead. A moment later it will have newly acquired the property defined by $\mathsf{PP}Kx$, then the one defined by $\mathsf{PPP}Kx$, and so on. The object undergoes a continuous change even though it keeps all of its ordinary properties constant.[79] Alas, tensed properties are as causally inert as anniversary properties. Since none of our accounts of time and change are committed to the tensed property comprehension schema, they are not committed to pure

[79]One problematic feature of this account is that an object would continue to change its tensed properties even after it has ceased to exist. We could rule out this possibility by restricting the comprehension schema to existing objects, $\exists X\, \forall x\, (Xx \leftrightarrow (\mathrm{E}!x \wedge \varphi(x)))$.

becoming in this sense, either. This includes tense primitivism, which does not require tensed properties any more than spatial views of time.

Throughout my discussion, I took it for granted that change is something that happens to material objects. This might seem to be an uncontroversial assumption, but not everybody accepts it. J. M. E. McTaggart attributes the changing-objects view to Bertrand Russell and claims that it is radically mistaken. He thinks that only *events* can undergo change:

> Mr Russell looks for change, not in the events in the time series, but in the entity to which those events happen, or of which they are states. If my poker, for example, is hot on a particular Monday, and never before or since, the event of the poker being hot does not change. But the poker changes, because there is a time when this event is happening to it, and a time when it is not happening to it. But this makes no change in the qualities of the poker. It is always a quality of that poker that it is one which is hot on that particular Monday. And it is always a quality of that poker that it is one which is not hot at any other time. Both of these qualities are true of it at any time—the time when it is hot and the time when it is cold. And therefore it seems to be erroneous to say that there is any change in the poker. The fact that it is hot at one point in a series and cold at other points cannot give change, if neither of these facts change—and neither of them does. (McTaggart 1927: bk. V, ch. 33, §315)

McTaggart goes on to develop a theory of time that regards tenses as properties of events, rather than as sentential operators. In the first place, this involves three basic tense properties *P*, *N*, and *F*:

$$
\begin{array}{lll}
Pe & \text{iff} & e \text{ is wholly past} \\
Ne & \text{iff} & e \text{ is present} \\
Fe & \text{iff} & e \text{ is wholly future}
\end{array}
$$

McTaggart also allows combinations of such event tenses, such as *PF*, which applies to events that were future, or *PFP*, which applies to events that were going to be past. The passage of time would consist in events undergoing change with regard to such tense properties. Every event is first *F*, then *N*, and finally *P*, and it would also undergo change with regard to more complex tense properties.

After spelling out this theory of tense properties, McTaggart's discussion takes a rather bizarre turn. In what is now known as *McTaggart's Paradox*, he goes on to argue that his own account of the tensed properties of events is in fact inconsistent (1927: ch. 33). He blames this on time itself, which he concludes to be unreal. Most contemporary authors resist this radical conclusion, either by putting the blame on McTaggart's theory of tense properties, or by denying that McTaggart's paradox is indeed paradoxical.[80] Instead of getting into the rather tedious details of this argument, I want to follow J. J. C. Smart and reject McTaggart's underlying assumption that change happens to events, rather than to material objects:

> The traffic light changed from green to amber and then it changed from amber to red. Here are two happenings, and these happenings are changes of [the] state of the traffic light. That is, *things* change, *events* happen. The traffic light changes, but the changing of the traffic light cannot be said to change. To say that it does or does not change is utter nonsense. Similarly, the traffic light neither does nor does not happen. We must also resist the temptation to misuse the word 'become'. The traffic light *was* green and *became* red, but the becoming did not become. Events happen, things become, and things do not just become, they become something or other. 'Become' is a transitive verb; if we start using it intransitively, we can expect nothing but trouble. (Smart 1949: 485–6)

We could ignore Smart's advice and treat events as changing, anyway, but that would not help much. Section 2.2 concluded that events are not the fundamental temporal entities. If that is right then McTaggart's account of changing events cannot be our fundamental theory of time, quite independently of any concerns about its consistency. Whatever tense properties events possess would have to be inherited from their constituent components. This leads us back to the question of whether ordinary material objects have tensed properties, and we already saw that there is no reason why a tense primitivist—or anyone else, for that matter—should accept such properties. All the philosophical problems that a theory of time needs to solve can be addressed without assuming that change is an essential feature of persistence.

[80] See, e.g. Dummett (1960), Lowe (1987; 1992; 1998), Mellor (1981a; 1998), Le Poidevin (1991), and Oaklander and Smith (1994: pt. III).

11

THE PROBLEM OF MOTION

So far, I pretended that the nature of space and time can be considered independently of one another. This simplifying assumption must now be given up. Questions about space and time are intimately connected by motion. Since an object moves if and only if it possesses different locations at different times, our metaphysics of motion must explain both what it is for an object to be located at a place and what it is for it to possess properties of location at a time. Instead of two separate issues, there is one *Problem of Motion* that subsumes the Problem of Change and the substantivalism/relationism debate about space.

11.1 Einstein's train example

To illustrate how the problem arises, Albert Einstein (1918: §3) presents the example of a train that moves at constant speed v relative to the embankment. Suppose Jane sits on the train and her friend Jack stands besides the railroad track. On the table in front of Jane, there is a stone. When asked about its state of motion, Jane tells us that, based on her observation, she judges the stone to be at rest. Jack gives a different answer to the same question. Peeking through the carriage window, he determines that the stone is moving at the same velocity v as the rest of the train. Jane and Jack both seem justified in their respective views, but the stone cannot be both at rest and moving. Part of this problem was already solved by Euclid's *Optics* (1883–1916: vol. VII, 110), long before trains were invented. Whether an object appears to be moving, Euclid explains, depends on whether or not the observer regards herself as moving. Jane is judging whether the stone moves relative to her and Jack is judging whether it is moving relative to him. They are no more contradicting one other than if Jane had said 'I am hungry', and Jack had objected 'No, I'm not'.

The same point can be made in a slightly more sophisticated way. To describe how objects move in space, physicists employ coordinate systems. By using appropriate measurement procedures, they assign each location in space a triple of numbers $\langle x, y, z\rangle$ as spatial coordinates and

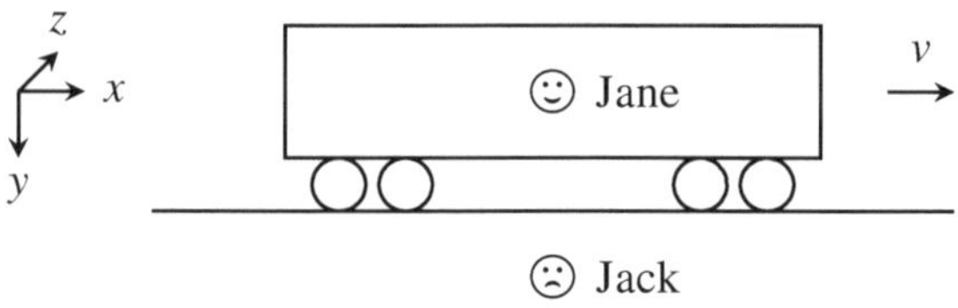

Figure 11.1. Einstein's train

each instance of time a number t as temporal coordinate (date). A *reference frame* is a combination of a spatial and a temporal coordinate system. There are infinitely many locations that we could choose as the origin of our spatial coordinate system, which is the point with coordinates $\langle 0, 0, 0 \rangle$, and there are infinitely many instants that we could pick as time zero of our temporal coordinate system. In the spatial case, we are also free to chose the orientation of the three coordinate axes, and in both the spatial and the temporal case the precise assignment of coordinates depends on which measuring devices we choose for setting up our coordinate systems. So there are infinitely many reference frames that we can choose from.

Given a choice of reference frame, an object's trajectory is a function from temporal to spatial coordinates that assigns to each time at which the object exists the location it occupies at that time. The mathematical form of a trajectory depends both on the motion of the object and on the choice of reference frame. The same object can therefore have quite different trajectories in different frames. Some frames describe the object as moving, by assigning it different spatial coordinates at different times, while others characterize it as being at rest, by always assigning it the same spatial coordinates. In our example, Jane and Jack tacitly assess the stone's state of motion relative to reference frames in which they themselves are at rest. Since Jane is claiming that the stone is at rest in her frame R and Jack is denying that it is at rest in his frame R^*, they are not asserting and denying the same claim.

Now suppose that Jane takes the stone and drops it out of the window. Relative to her frame R, it falls on a straight line down to the railroad embankment. Its trajectory looks different in Jack's frame. Since R^* moves relative to R, the stone describes a parabolic curve in R^*. This apparent disagreement about the stone's motion can be resolved by translating Jane's description into Jack's reference frame by means

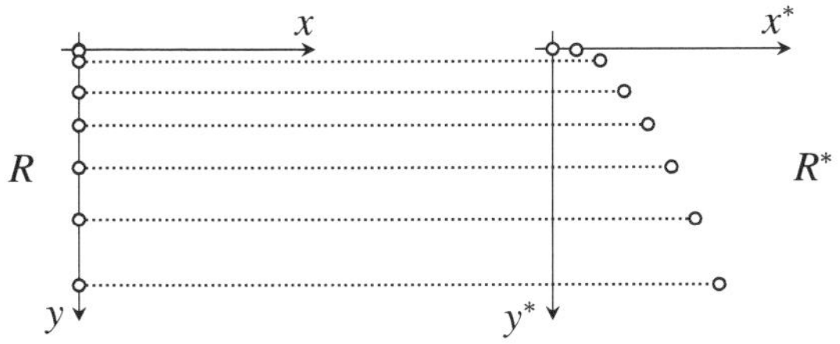

Figure 11.2. The falling stone

of a *coordinate transformation*. To keep the mathematical details simple, suppose that Jane and Jack use the same type of measuring devices to set up their reference frames, and suppose further that both arrange their coordinate systems so that in Fig. 11.1 the x-axis is pointing to the right, parallel to the tracks, the y-axis is pointing to the bottom of the page, and the z-axis is pointing away from the reader. Let us also suppose that both assign the temporal coordinate zero to the precise instant at which their spatial coordinate systems coincide. If we use x, y, z, and t for the coordinates in R and assign the stone spatial coordinates $\langle 0, 0, 0 \rangle$ at time 0 then its trajectory in Jane's frame is:

$$x(t) = 0 \qquad y(t) = \frac{1}{2}gt^2 \qquad z(t) = 0$$

Here g is a constant determined by the local strength of the Earth's gravitational field. Translating this into Jack's frame R^* appears to be a straightforward matter. At each time t, we need to add to Jane's x-coordinates the distance vt that her frame R has moved relative to Jack's frame R^*. Since the train is only moving in the x-direction, the other coordinates remain unchanged. If we use x^*, y^*, z^*, and t^* for the coordinates in R^* then we get what is known as a *Galileo transformation*:

$$x^* = x + vt \qquad y^* = y \qquad z^* = z \qquad t^* = t$$

If we apply this transformation to the trajectory of the stone in R then we get the parabolic curve observed by Jack in R^*:

$$x^*(t^*) = vt^* \qquad y^*(t^*) = \frac{1}{2}gt^{*2} \qquad z^*(t^*) = 0$$

Jane and Jack thus appear to be in perfect agreement about the motion of the stone; they merely describe it from different perspectives.

Everybody agrees that we can only speak of motion relative to a reference frame. Opinions begin to diverge at the next step, when we turn to the task of explaining the contribution of reference frames in time-relative attributions of location:

Object a has spatial coordinates $\langle x, y, z\rangle$ at time t relative to frame R

The *Problem of Motion* is the task of determining the logical form of such statements. This problem used to receive little attention because it was thought to consist of two independent sub-problems. A reference frame can be regarded as a pair $R = \langle R_s, R_t\rangle$, where R_s encodes the prescriptions for assigning spatial coordinates and R_t those for temporal coordinates. The Problem of Motion thus divides into two parts:

$$\underbrace{\underbrace{[\text{Object } a \text{ has coordinates } \langle x, y, z\rangle \text{ relative to } R_s]}_{\text{(i)}} \text{ at } t \text{ relative to } R_t}_{\text{(ii)}}$$

Part (i) concerns the dispute between spatial substantivalists and relationists about how to understand spatial coordinates. Spatial substantivalists regard coordinates as names for metaphysically basic points on a three-dimensional spatial manifold $\mathcal{M}$ while relationists about space take them to be nothing more than labels for the positions in the order that spatial relations impose on material objects. Part (ii) raises the question of what it is for an object to possess a property at a time, which is just the familiar Problem of Change. We seem to have two independent problems, one about space and one about time, that jointly exhaust what is problematic about the Problem of Motion.

11.2 The relativity of simultaneity

The divide-and-conquer approach to the Problem of Motion takes it for granted that spatial and temporal coordinates can be assigned independently of one another. It was Einstein's seminal insight to realize that this is not only an *empirical* assumption, but that it is also *false*. This insight was triggered by the unexpected discovery, in the late nineteenth century, that the speed of light in a vacuum is the same in all reference frames. Given two frames R and R^*, light propagates at the same speed of $c = 299,792,458$ m/s in both frames, even if one is moving relative to the other. This discovery has far-reaching theoretical consequences.

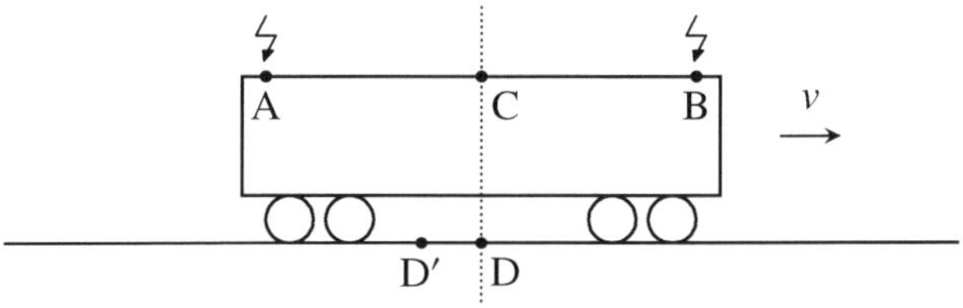

Figure 11.3. The relativity of simultaneity

Consider the task of determining whether two events occur simultaneously. If the events occur at the same place then an observer who is located there could observe directly whether they happen at the same time, or whether one happens before the other. This is not possible for events that happen at different locations; for them, simultaneity has to be inferred from other observations. Take our moving train example and suppose that Jane is sitting at a location C that is exactly half-way between two points A and B on the train. Let us assume that Jane knows about the relative locations of A, B, and C because she previously measured the distances between them with the measuring devices she used to set up the spatial part of her reference frame.

Next, suppose that lightning strikes at A and B, and that Jane at C observes the light flashes emitted from these two places at exactly the same time. Since Jane knows that A and B are equidistant from C, and since she also knows that the speed of light is constant, she concludes that A and B were struck simultaneously. Given her knowledge about A, B, and C and the propagation of light, she is able to infer from the directly observed simultaneity of two co-located events (the arrival of the two light flashes at C) that the spatially separated bolts of lightning at A and B happened simultaneously.

Let us also assume that Jane is passing Jack at position D at the very moment when, according to Jane, lightning strikes A and B. Jack does not receive the light flashes from A and B at the same time, which Jane might explain as follows. Relative to her frame *R*, Jack and the light source at A are moving towards one another and the flash from A reaches Jack at some point D′. Since the distance from A to D′ is shorter than the distance from D′ to B, and since the two light rays travel at the same speed, Jack sees the flash from A before he sees the flash from B.

So far, so good. Now consider how Jack might describe the events from his perspective. The seemingly obvious way of generating such a description is by using the Galileo transformation to translate Jane's description into Jack's frame R^*. This would tell us that, relative to R^*, the flash from A moves at a speed $c + v$ that is equal to the sum of the speed of light c in Jane's frame R and the speed v with which her frame moves relative to Jack's. Since the light flash from B moves in the opposite direction as R, it propagates at the smaller speed $c - v$.

This might yield the expected result that Jack receives the flash from A before the flash from B, but this explanation cannot be right if the speed of light is the same in all reference frames. If Jack actually measured the velocities of the two light rays then he would find that both move with the same velocity c relative to his reference frame R^*, rather than one moving with $c + v$ and the other with $c - v$. This means that the Galileo transformations, which initially looked like simple mathematical truths, are in fact *false*. If the speed of light is the same in all reference frames then the Galileo transformations cannot be the correct way of translating coordinates from one frame to another.

Since velocity is the ratio of distance travelled and time elapsed, the speed of light can only be constant across reference frames if either distances or durations depend on the choice of frame. The correct answer, it turns out, is a bit of both. Finding the appropriate coordinate transformation is an exercise in reverse engineering. The task is to figure out how lengths and durations must depend on the reference frame so that light ends up having the same velocity c in all frames. The solution to this mathematical problem is that we need to translate Jane's coordinates into Jack's by using the so-called *Lorentz transformation*:[81]

$$x^* = \frac{x + vt}{\sqrt{1 - \frac{v^2}{c^2}}} \qquad y^* = y \qquad z^* = z \qquad t^* = \frac{t + \frac{v}{c^2}x}{\sqrt{1 - \frac{v^2}{c^2}}}$$

(Note that these equations approximate the Galileo transformation for small values of v, for which v/c^2 is effectively zero. We only get an observable effect if the velocity v is close to the speed of light, which is *very* fast. In everyday life, the differences between the two transformations are so tiny that they fall well within the margin of error of any measurement. To confirm the predictions of the theory of relativity, we need

[81] The derivation can be found in Einstein (1918: 78–82) or in any textbook on the theory of relativity, such as Taylor (1975: sec. 3.3) or Giulini (2005: sec. 3.2).

objects that move much faster than the train in Einstein's thought experiment. One popular choice is the earth itself, whose velocity around the sun is about a tenth of the speed of light.)

The Lorentz transformations assume that the velocity v with which one reference frame moves relative to another is strictly smaller than the speed of light c. Reference frames with $v = c$ cannot be permitted without wreaking mathematical havoc with the Lorentz transformations, for if v were identical to c then the denominator $\sqrt{1 - v^2/c^2}$ would equal zero. But it is not immediately clear *why* the case $v = c$ needs to be ruled out. By continuing to exert an external force on the train, could we not, at least in principle, accelerate it up to the speed of light? If so, then our theory ought to predict something about the measurements of distances and durations on such a train. One might also wonder what is so special about *light*, rather than, say, the speed of sound or water waves, that its velocity gets to play such a prominent role in our account of space, time, and motion.

The answer to both questions is provided by relativistic dynamics. If the laws of physics are to take the same form in Jane's reference frame as they do in Jack's, then we must abandon classical Newtonian mechanics. What we get instead is a theory in which an object's mass m depends on its velocity v in the reference frame under consideration:

$$m = \frac{m_0}{\sqrt{1 - \frac{v^2}{c^2}}}$$

Here m_0 is the *rest mass* of the object, which determines how it accelerates in response to external forces when it has zero velocity. Since the velocity of an object depends on the reference frame, the same object has a different relativistic mass m in different reference frames; only its rest mass m_0 is always the same. Given the way relativistic mass depends on velocity, it becomes rapidly larger the closer v gets to the speed of light c. Since we would require more and more energy to accelerate the object even further, it is physically impossible to accelerate any material object to the speed of light because doing so would require an infinite amount of energy. The only exception to this rule are objects whose rest mass m_0 is zero, which are only permitted to move at velocity c. This includes light particles (photons) and the gluons that mediate interactions between the particles that compose the nuclei of atoms. What is special about the speed of light is not that it is the speed of *light*, but that it is the velocity of objects with rest mass zero.

For current purposes, the most important feature of the Lorentz transformations is what they tell us about the simultaneity of spatially separated events. In Jane's reference frame *R*, the bolts of lightning at A and B strike simultaneously, and both points are equidistant from C. In Jack's reference frame R^*, the two bolts of lightning do not strike A and B at the same time, nor is the distance between *A* and *D* the same as the distance between *B* and *D*.

One might try to blame this result on the assumption that the events at A and B happen simultaneously if and only if an observer placed exactly half-way between them sees their flashes at the same time. Instead, one might propose placing two synchronized clocks at the two locations and to use them to determine the times at which the two events take place. But Einstein (1956: 30) points out that this just raises the question of whether there is a frame-independent way of synchronizing clocks that are placed at different locations. We could synchronize two clocks at D and then move them to A and B, respectively, but that would only help if we were assured that the two clocks continue to tick at the same rate when they are on the moving train, and the theory of relativity entails that they do not. Since similar considerations apply to other methods for assigning dates to spatially separated events, Einstein concludes that there is no frame-independent notion of simultaneity:[82]

> The concept does not exist for the physicist until he has the possibility of discovering whether or not it is fulfilled in an actual case. We thus require a definition of simultaneity such that this definition supplies us with the method by means of which, in the present case, he can decide by experiment whether or not both the bolts of lightning occurred simultaneously. As long as this requirement is not satisfied, I allow myself to be deceived as a physicist (and of course the same applies if I am not a physicist), when I imagine that I am able to attach a meaning to the statement of simultaneity. (Einstein 1918: ch. VIII)

Few contemporary philosophers will endorse the verificationist theory of meaning that Einstein seems to take for granted in this passage, but

[82]This is the standard reading of the Lorentz transformations, but not the only one. Lorentz (1913), after whom the translations are named, accepts absolute facts about simultaneity and tries to account for the constancy of the speed of light by postulating a contraction of objects that move relative to the *ether*, which is the medium in which electromagnetic waves are said to propagate. On this view, durations and distances do not change with velocity, material objects do. This yields an observationally indistinguishable theory, but one that has larger ontological commitments than Einstein's theory. See also Schlick (1915: sec. IV), van Fraassen (1970: sec. V.2), and Bell (2004).

rejecting this assumption would not get rid of the problem. Quite independently of our views about the nature of meaning, we saw in Section 7.3 that spatial and temporal coordinate systems partially depend on our scientific theories. Only the simultaneity of events that occur at the location of the observer can be the subject of direct observation. Judgements about the simultaneity of spatially separated events must rely on an inference that appeals to some physical theory. The novel feature of the theory of relativity is that it does not support any frame-independent way of making such judgements.

11.3 Minkowski spacetime

According to the Lorentz transformations, the same method for assigning temporal coordinates can yield different results about the simultaneity of events when it is combined with different methods for assigning spatial coordinates. Even though Jane and Jack apply the same methods for measuring distances and durations to the same events, they come up with different results. The Problem of Motion can still be divided into two sub-problems, but if the methods for assigning spatial and temporal coordinates depend on one another then these sub-problems are not independent of one another. Rather than treat the two issues separately, we must find a joint solution to the Problem of Change and the substantivalism/relationism debate about space.

The standard solution is due to Hermann Minkowski, who suggests that we treat space and time as mere facets of a four-dimensional *spacetime*.[83] The Lorentz transformations tell us that neither spatial nor temporal distances remain the same when we move from one reference frame to another. But since the speed of light is the same in all reference frames, every change in spatial distances must be compensated by a change in temporal distances and the so-called *spacetime metric* g is easily seen to be invariant under Lorentz transformations:

$$g(s, s') = \underbrace{(x' - x)^2 + (y' - y)^2 + (z' - z)^2}_{\text{(spatial distance)}^2} - \underbrace{c^2(t' - t)^2}_{c^2\text{(duration)}^2}$$

Here $s = \langle x, y, z, t\rangle$ and $s' = \langle x', y', z', t'\rangle$ are any two spacetime points.

[83] 'Von Stund an sollen Raum und Zeit für sich völlig zu Schatten herabsinken, und nur eine Art Union der beiden soll Selbstständigkeit bewahren' (Minkowski 1909: 104). Some authors believe that the spacetime idea is really due to Poincaré (1906). More details of the history of spacetime proposal can be found in Einstein (1921; 1929; 1956), Jammer (1954), Grünbaum (1955; 1962), Stein (1977), and Friedman (2002).

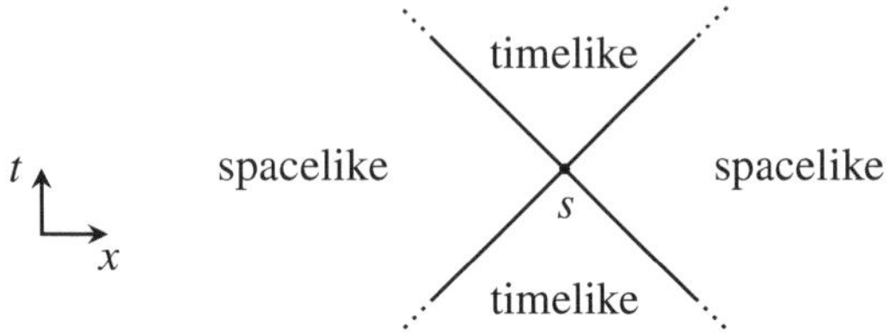

Figure 11.4. The lightcone

The spacetime metric allows us to make a number of physically important distinctions. Suppose we plot the time coordinate on the vertical axis and suppress all spatial coordinates except for the x-coordinate. Two points $s = \langle x, t \rangle$ and $s' = \langle x', t' \rangle$ on this two-dimensional spacetime are said to be at a *lightlike* distance from one another just in case:

$$g(s, s') = (x' - x)^2 - c^2(t' - t)^2 = 0$$

If we simplify matters further by choosing our units of length and duration so that $c = 1$ then this is equivalent to the condition $x' - x = \pm(t' - t)$, which is satisfied by all the points on the diagonals emanating from s. The collection of all of these points is called the *lightcone* of s. Points outside the lightcone satisfy $g(s, s') > 0$ and are said to be at a *spacelike* distance from s, while those inside the cone satisfy $g(s, s') < 0$ and are said to be at a *timelike* distance from s.

Since no object can move faster than the speed of light, what is happening at s can only be causally influenced by what is taking place on or within its backward lightcone. If a point s' is a spacelike distance from s then either the temporal coordinate of s' is already later than that of s, or else the spatial distance between x and x' is too large for any signal to cover it within the time $t' - t$. Even a light-ray sent off from s' will only get to x after t. What is happening at s' can thus have no causal impact on what is going on at s. The points on the backward lightcone itself are just close enough so that a light-ray sent off from x' at time t' arrives at x exactly at time t. A light-ray sent from within the backward lightcone already passes x before t, but a slower means of interaction might still affect s. Similarly, what is happening at s can have no causal impact on spacetime points outside its forward lightcone because they are too far in space or too close in time. Points on the forward lightcone itself can be reached by a light-ray from s.

The spacetime view also offers a nice explanation of why there are no frame-independent relations of simultaneity. Relative to a reference frame, a spacetime point s is simultaneous with a point s' just in case the two get assigned the same temporal coordinate. Otherwise, the point with the larger temporal coordinate is later relative to that frame. Because lightcones are invariant under Lorentz transformations, any spacetime point in the forward lightcone of s is later than s in all reference frames, and any spacetime point in the backward lightcone is earlier than s in all reference frames. This does not extend to spatially separated points. Given a point s' outside the lightcone of s, we can find one reference frame in which s' is later than s, a second frame in which s and s' are simultaneous, and a third frame in which s' is earlier than s. The temporal order of spacelike separated spacetime points always depends on the choice of reference frame. To ask, of two spatially separated events, whether they are simultaneous is like asking of two points in space whether they have the same x-coordinate. Such questions make sense relative to a chosen frame of references, but there are no frame-independent relations of simultaneity any more than there are frame-independent relations of being 'co-x-ed'.

As in the case of space alone, one could envisage either a relationist or substantivalist account of spacetime. Spacetime substantivalism postulates a four dimensional manifold $\mathcal{S}$ that exists independently of its contents. A reference frame R defines a map f_R from quadruples of real numbers to points on $\mathcal{S}$ and an event e occurs at place $\langle x, y, z \rangle$ at time t just in case e is spatiotemporally located at $f_R(x, y, z, t)$ on $\mathcal{S}$. Relationist accounts of spacetime are a little bit more complicated. To avoid running afoul of relativity theory, we would need to make sure that our theory of spatial and temporal relations is weak enough so that we cannot define a frame-independent simultaneity relation in terms of our primitives. One such proposal is Alfred Robb's (1914; 1921) *theory of conical orders*, which restricts the temporal precedence relation so that an event only precedes the events in its forward lightcone.

12

RELATIVITY WITHOUT SPACETIME

EVERYBODY agrees that the spacetime picture is very helpful in representing physical processes and the laws that govern them in a frame-independent manner. What is less obvious is whether we should also regard it as a literal account of the fundamental nature of space and time. It is widely thought that theory of relativity makes such a view unavoidable because it prohibits us from splitting spacetime into a spatial and a temporal component. I want to argue that this *inseparability argument* is in fact inconclusive and that there is room for a rival view that combines spatial substantivalism with tense primitivism. Such a hybrid account does justice to the significant differences between space and time and respects the empirical discoveries of relativity theory by recovering spacetime on a higher level of abstraction.

12.1 The inseparability argument

If we wanted to, we could formulate not only the theory of relativity, but also classical Newtonian mechanics on a four-dimensional spacetime manifold. The difference is that the spatial and temporal component of such a Newtonian spacetime are easily pried apart. Given any spacetime point, its *simultaneity hypersurface* consists of all spacetime points that are simultaneous with that point. Such a hypersurface is a three-dimensional subspace that splits spacetime in two. On one side are points that are later than all the points on the hypersurface, and on the other side are all the earlier points. By constructing simultaneity hypersurfaces for all points in this way, we obtain a *foliation*, which is an exhaustive division of spacetime into non-overlapping hypersurfaces. This allows us to identify classical spacetime with the Cartesian product $\mathcal{M}\times\mathcal{T}$ of a three-dimensional spatial manifold $\mathcal{M}$ and a one-dimensional temporal manifold $\mathcal{T}$. Newtonian spacetime can thus be thought of as a three-dimensional spatial manifold that evolves in time.

The special theory of relativity lacks the absolute notion of simultaneity on which this construction is based. Given a choice of reference frame that assigns dates and locations to all events, spacetime points with the same temporal coordinate form a *spacelike hypersurface*. This is a three-dimensional subspace on which any two points are at a spacelike distance from one another. This does yield a foliation of spacetime, but different choices of reference frame carve up spacetime in different ways. It is only relative to a reference frame that we can distinguish the temporal coordinates of a spacetime point from its spatial coordinates. Unlike in the case of Newtonian mechanics, there is no privileged foliation that permits a frame-independent decomposition of a relativistic spacetime into a spatial and a temporal component.[84]

Space and time appear to be even more intimately intertwined in the general theory of relativity, which claims that gravitational interactions are mediated by a distortion of the geometry of spacetime. The presence of a mass causes spacetime to curve and it is this curvature that accounts for the gravitational effects on masses located elsewhere. The spacetime metric g in general relativity is not a constant, as it is in the special theory, but depends on the distribution of mass. Locally, the spacetimes of general relativity look like Minkowski space with the 'flat' metric from p. 119, but their global structure can be quite different. In this respect, curved four-dimensional spacetimes are similar to the curved two-dimensional surface of a sphere. In a small neighbourhood of any point, the curved surface can be approximated by a flat two-dimensional tangent plane. This is what underlies the local approximation of areas on the surface of the earth by flat maps. The farther we move away from the tangent point, the more noticeable the curvature of the sphere becomes. To compensate for this, we need an atlas of maps that jointly cover the entire surface. Likewise, we can use small regions of flat Minkowski space to form an atlas of a curved spacetime.

The curvature of spacetime is usually quite benign, but there are models of general relativity, discovered by Kurt Gödel, in which spacetime admits no foliation at all.[85] In a small neighbourhood of any point, we can establish a coordinate system that locally orders events into simultaneity classes, but the spacetime is so twisted that these local definitions cannot be patched together to yield global spacelike hypersurfaces. In such cases, there is not even a frame-dependent way of splitting spacetime into a spatial and temporal part. The inseparability argument concludes that relativistic spacetime is an inseparable amalgam, and that any attempt at treating space and time separately is bound to result in

[84] See Einstein (1956: 31) and Russell (1969: ch. V).

[85] Gödel (1949a;b); see also Earman (1970: 260), Stein (1970), and Savitt (2005).

failure. Since spacetime is not a marriage of equals, but an annexation of time by space, this would mean that the considerations against spatial views of time presented earlier get overruled by our best physics.

If this is correct then a further conclusion of philosophical interest follows. Minkowski offers us a spacetime occupied by *events*, and we saw in Section 2.2 that events are not the basic temporal entities. Events are metaphysically complex entities that inherit their temporal features from their constituent components. As it stands, Minkowski's proposal therefore fails to provide a solution to the Problem of Motion, which demands an account of what it is for a *material object* to have a spatial location at a time. One could of course say that an object is located at point $\langle x, y, z \rangle$ at time t just in case it participates in an event that occurs at the corresponding point in spacetime, but that would lead us back to the intractable problem of finding an account of participation that does not already assume the notion of spatiotemporal location. If we want to hold on to spacetime as a fundamental entity then we probably need to endorse the *perdurantist* view that objects persist by having different temporal parts at different times.[86] This would make material objects more similar to events, by ascribing them both spatial and temporal extension. If spacetime is populated by such four-dimensional material objects then we could solve the Problem of Motion by saying that object a has spatial coordinates $\langle x, y, z \rangle$ at time t relative to reference frame R just in case it has a spatiotemporal part a-at-$f_R(x, y, z, t)$. Here f_R is a (perhaps only partially defined) function from coordinates to points on the spacetime manifold given by R.

The theory of relativity thus appears to settle all philosophical questions about time. It mandates that we treat time on the model of space, it speaks in favour of temporal parts, and it also seems to rule out a metaphysically physically privileged role for the present moment. If we accept Newton's rotating bucket argument and Field's field argument then our physics also settles the relationism/substantivalism dispute in favour of a substantivalist spacetime. Hilary Putnam concludes that 'there are [no] longer any *philosophical* problems about time; there is only the physical problem of determining the exact physical geometry of the four-dimensional continuum that we inhabit' (1967: 247). As a scientifically-minded philosopher, I would be happy to accept the conclusion that all questions in the philosophy of time are really just scientific questions, but I do not think that the inseparability argument

[86] See Smart (1972), Sider (2001: sec. 4.4), Sattig (2006: ch. 2), and Balashov (2010).

succeeds in establishing this thesis. The problem lies with the unquestioned assumption that treating space and time differently requires a separable spacetime, which I believe to be false.

Contrary to a widely held view, the Lorentz transformations do not show that space and time can somehow be converted into one other, and must therefore be made of the same material. When transforming coordinates from one reference frame to another, the new time coordinate of an event might be a function of its old temporal coordinate *and* its spatial coordinates, but the theory of relativity does not permit the following scenario, imagined by Donald Williams:

> One four-dimensional area of the manifold [is] slewed around at right angles to the rest, so that the time order of that area ... runs parallel with a spatial order in its environment. It is conceivable, indeed, that a single whole human life should lie thwartwise of the manifold, with its belly plump in time, its birth at the east and its death in the west, and its conscious stream perhaps running alongside somebody's garden path. (Williams 1951: 468)

We can change the spatial coordinates of events by rotating, re-scaling, or moving our spatial coordinate axes, and we can independently change their temporal coordinates by changing the origin or units of our time axis. The only way of 'mixing' spatial and temporal coordinates is by shifting to a moving reference frame. But since the spacetime metric g is invariant under Lorentz transformations, so is the distinction between spacelike and timelike distances. There is no Lorentz transformation that would effect the strange rotation envisaged by Williams, and turn a sequence of timelike separated events, such as a human life, into a sequence of spacelike separated ones. While one can rotate a cube in space to reveal a spatial side that was previously hidden, one cannot rotate an event in Minkowski space to show its temporal side.[87]

On its own, the theory of relativity does not prohibit a frame-independent simultaneity relation between events. One can consistently supplement the theory with the stipulation of a privileged reference frame that captures the truth about simultaneity.[88] Any two events that are assigned the same date by this special frame would then be simultaneous

[87]Even if we accept spacetime as a fundamental entity, there are thus still residual differences between space and time. These differences largely stem from the fact that spatial and temporal coordinates enter into the calculation of spacetime distances with different signs. See Skow (2007) for a more detailed discussion of these issues.

[88]Tooley (1997: ch. 11) exploits this option to reconcile his growing block view with the theory of relativity. See also the references in footnote 82.

simpliciter, but there would be no way of determining, empirically or otherwise, which of the infinitely many reference frames is privileged in this way. To rule out such a gratuitous addition to our metaphysics, we need a parsimony principle that prohibits us from postulating more spatial and temporal structure than is needed to account for our best physical theories. Suppose we accept some such principle. We could then count it as one of the consequences of the theory of relativity that there are no frame-independent simultaneity relations.

This allows us to derive results that appear to support the inseparability argument. Suppose we tried to be substantivalists about space and time separately, by postulating a three-dimensional spatial manifold $\mathcal{M}$ and a one-dimensional temporal manifold $\mathcal{T}$. Their Cartesian product $\mathcal{M} \times \mathcal{T}$ would then give us a four-dimensional spacetime manifold, but we could define a frame-independent notion of simultaneity in terms of temporal co-location on $\mathcal{T}$. If we want to be substantivalists about space and time then we must be spacetime substantivalists. Similarly, suppose we tried to be relationists about space and time separately, by postulating enough spatial relations to permit an abstraction of locations from spatially ordered material objects, and enough temporal relations to permit an abstraction of instants of time as classes of temporally ordered events. By taking the Cartesian product of the sets of these locations and instants, we get an abstract, four-dimensional spacetime. In this case, we could define a frame-independent notion of simultaneity by regarding two events as simultaneous whenever they are elements of the same instant of time. If we want to be relationists about space and time then we must be relationists about spacetime.

These are important constraints on our accounts of space and time, but they fall short of establishing the conclusion of the inseparability argument. All they show is that we must endorse a spacetime view if we are *already* committed to treating space and time alike, by adopting either a substantivalist or relationist view about both. Since they do not rule out other combinations of accounts of space and time, they provide no independent reason for thinking that time is like space.

12.2 Hybrid views

Suppose we accept Newton's rotating bucket argument and Field's field argument in favour of spatial substantivalism. We saw in Section 3.2 that neither argument carries over to the temporal case, so there is no particular reason why we should also endorse temporal substantival-

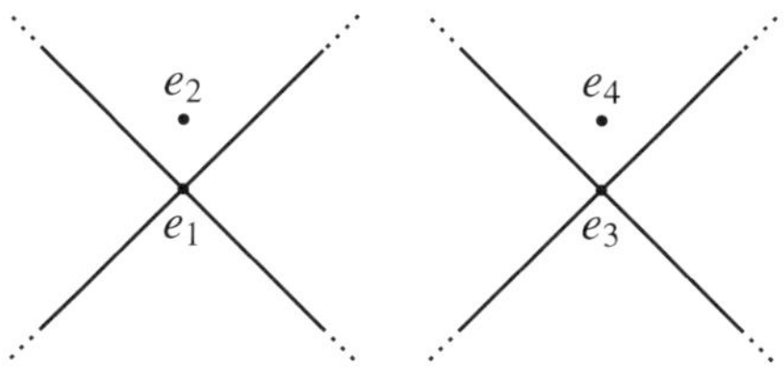

Figure 12.1. Four events

ism. Since spatial substantivalism already has the resources to account for inertial forces and free fields, we are at liberty to adopt a different view about the nature of time. One option is to combine spatial substantivalism with relationism about time, the other to supplement it with tense primitivism. Contrary to what the inseparability argument suggests, both of these hybrid views can be made to respect the proscription against frame-independent simultaneity relations.

When attempting to combine spatial substantivalism with relationism about time, we need to make sure that our theory of temporal relations is weak enough so that we cannot define any frame-independent simultaneity relation in terms of our primitives. One way of doing this is by adopting Robb's theory of conical orders mentioned on p. 121. This theory is usually thought of as a relational theory of spacetime, but the same strategy helps with the current project. Consider the three conditions on temporal overlap ($\circ$) and complete temporal precedence ($\ll$) from p. 9 that jointly ensure that the earlier-than relation on Russellian instances is irreflexive, transitive, and comparable:

R1 $\forall e \neg e \ll e$

R2 $\forall e_1 \forall e_2 \forall e_3 \forall e_4 ((e_1 \ll e_2 \wedge e_2 \circ e_3 \wedge e_3 \ll e_4) \rightarrow e_1 \ll e_4)$

R3 $\forall e_1 \forall e_2 (e_1 \circ e_2 \leftrightarrow \neg(e_1 \ll e_2 \vee e_2 \ll e_1))$

We cannot retain this theory of temporal relations if we want to follow Robb and restrict $\ll$ to forward lightcones. Take four instantaneous events e_1, e_2, e_3, and e_4 that are placed in spacetime as depicted in Fig. 12.1. On Robb's reading of the precedence relation, neither $e_2 \ll e_3$ nor $e_3 \ll e_2$ holds because these events occur at a spacelike distance from one another. With R3, this entails $e_2 \circ e_3$. Since e_2 is in the forward lightcone of e_2 and e_4 is in the forward lightcone of e_3, we also get $e_1 \ll e_2$ and $e_3 \ll e_4$, and thus $e_1 \ll e_4$ with R2. Since e_4 is not in the forward lightcone of e_1, this violates Robb's reading of $\ll$.

One way of adjusting our theory is to keep R1 and R2 and give up R3. In this case, we could no longer use R3 to define temporal overlap in terms of precedence. Instead, we would define overlap operationally, in terms of our choice of reference frame. Given a convention for assigning temporal coordinates, we can map each event to an interval of real numbers that contains all the dates at which the event occurs. Two events overlap relative to such a choice of frame just in case their associated intervals of real numbers have a non-empty intersection.[89] Depending on which frame we pick, we get different verdicts about overlap, and different Russellian time series. If $e_1 \ll e_2$ then e_1 is earlier than e_2 in all such time series, but if neither $e_1 \ll e_2$ nor $e_2 \ll e_1$ holds then their order depends on the choice of reference frame. In some frames, e_1 and e_2 overlap; in others, one of them precedes the other. There are frame-independent temporal relations, which include $\ll$ and everything that can be defined in terms of it, but they need not suffice to determine a unique temporal order of events. We have a relationist account of time without a frame-independent notion of simultaneity.

This hybrid view takes spatial coordinates to name points on the three-dimensional spatial manifold $\mathcal{M}$ and temporal coordinates to pick out Russellian instants that are defined, as on p. 9, in terms of the precedence relation $\ll$ and the frame-dependent overlap relation $\circ$. The Cartesian product of $\mathcal{M}$ and the set of these instants yields what one might call a *frame-relative* spacetime. Since the details of the construction depend on the choice of reference frame, there are many different frame-relative spacetimes of this type.

A coordinate-free spacetime can be obtained on a higher level of abstraction. A Lorentz transformation from one frame-relative spacetime to another is an *isomorphism* just in case it maps every feature of one spacetime onto a corresponding feature of the other. For example, if the temperature at a point in the first frame-relative spacetime is 125 °C then that point must be mapped to a point in the second spacetime with the same temperature.[90] Two spacetime points are said to be *equivalent* just in case there is an isomorphism that maps one to the other. The *equiva-*

[89]Note that overlap is not the same as simultaneity. Unlike simultaneity, overlap is not transitive. Consider the events e_1, e_2, and e_5 in Fig. 1.2 on p. 3. Both $e_1 \circ e_5$ and $e_5 \circ e_2$ are true, but $e_1 \circ e_2$ is false. The event e_1 overlaps with an early stage of e_5 and it wholly precedes e_2, which overlaps with a later stage of e_5.

[90]It is a non-trivial matter to extend the notion of a 'corresponding feature' to vector and tensor quantities, but the details of this do not matter here.

lence class of a point contains all points in all frame-relative spacetimes to which it is equivalent. Since each frame-relative spacetime is isomorphic to itself by way of the identity transformation, the equivalence class of a point always contains the point itself. Given a frame-relative spacetime, the associated coordinate-free spacetime then consists of all the equivalence classes of its spacetime points.

Such coordinate-free spacetimes do not admit privileged foliations. Because the frame-relative spacetimes out of which they are constructed disagree on the temporal order of spacelike separated events, there is no frame-independent way of splitting a coordinate-free spacetime into a spatial and a temporal component. Contrary to what the inseparability argument suggests, though, this is *not* an indication of a fundamental similarity between space and time, which this hybrid proposal expressly denies. Of course, this combination of spatial substantivalism and relationism about time does not avoid the general problem, discussed in Section 2.2, that events are not the basic temporal entities. The point I want to make here is that the theory of relativity does not give rise to any additional difficulties. If event-relationism about time were tenable in its own right then we could combine it with spatial substantivalism in a way that is compatible with the rejection of absolute simultaneity, and without postulating a separable spacetime.

The view I want to promote combines spatial substantivalism with my own modal theory of time. While relationists accommodate the discoveries of relativity theory by weakening their theory of temporal relations, tense primitivists begin by noting that the truth-values of tensed sentences depends on the choice of reference frame. Tense operators are devices for drawing simple temporal coordinate lines across space, and thus fall within the scope of the general conclusion of Section 7.3 that coordinate systems cannot outstrip the spatial and temporal measurements admitted by our best physical theories. Since the theory of relativity discovered that distances and durations depend on the choice of reference frame, the same holds for the basic temporal distinctions we can make with the help of tense operators. The truth-values of tensed sentences depends on the choice of reference frame. This does not mean that we made an empirical discovery about the meanings of P and F. Had we been attentive to the theory-dependence of coordinate systems then we would have realized all along that it is an empirical question whether there are any facts about the simultaneity of spatially separated events that are independent of our choice of reference frame.

We can use this insight to reconcile tense primitivism with the theory of relativity. Given a choice of reference frame, we construct an ersatz time series in the usual way, as discussed in Chapter 6. The Cartesian product of the spatial manifold $\mathcal{M}$ with this ersatz time series yields a frame-relative spacetime that we can populate with events that are constructed out of times, places, material objects, and properties.[91] There are as many such frame-relative spacetimes as there are reference frames, and no presumption that any one of them is privileged over the others. As in the relationist case, we then construct coordinate-free spacetimes by taking equivalence classes of points in frame-relative spacetimes. This account offers a way of reconciling the best metaphysics of time (i.e. tense primitivism) with our best physical theory of motion. Neither spacetime nor events would count as fundamental entities, but we recover both at a higher level of abstraction. So there is no need for physicists to give up their familiar spacetime talk. What is being proposed is a way of making metaphysical sense of spacetime, not a philosophically-inspired reform project for physics.

This view relies on some of the negative results established earlier. For example, if *presentists* were right that there is a non-trivial sense in which only present objects exist then there would be a privileged reference frame that correctly counts all existing objects as present. This would yield a frame-independent notion of simultaneity and thus conflict with the theory of relativity. To ward off this challenge, we need the result of Chapter 9 that tense primitivism is not a presentist thesis. We also need the conclusion of Chapter 5 that tense primitivists can remain agnostic about what sort of propositions or facts there are. If our account of time were committed to *tensed facts* then there would be a privileged frame that gets all the present facts right.

To extend this account to the curved spacetimes of general relativity, we need to take note of the empirical discovery that distances and durations depend on the distribution of mass. The mathematical details of the construction of coordinate-free spacetimes are a little bit more complicated in this case, but the underlying idea is the same as before. However, there remains the question of what to say about Gödel's examples of spacetimes that admit no global spacelike hypersurfaces. One option is to simply reject these examples, by following D. H. Mellor's counsel against the 'pernicious fallacy that ... anything which is phys-

[91] The details of the construction do not matter for current purposes; any of the Strawsonian views of eventhood listed in Section 2.2 would do.

ically possible must be possible in fact' (1998: 127). But suppose we accept them as presenting genuine possibilities that need to be accommodated by our theory of space and time. What I would say about such cases is that they realize a possibility that was always admitted by the theory-dependence of spatial and temporal coordinate systems. There is no a priori guarantee that our scientific theories will always support a way of aligning the dates of events across space. Reduced to the level of tense operators, this means that there might only be a local way of understanding tensed sentences, and thus only a local way of abstracting times from them. In such a case, there would be something irreducibly 'patchy' about the temporal aspect of reality. This might be a rather peculiar conclusion to arrive at, but Gödel's spacetimes *are* peculiar.

12.3 Symmetries and structure

Some philosophers might complain that the combination of spatial substantivalism and tense primitivism provides space and time with more structure than is warranted by our best theories of motion. Let a spacetime be a set of points equipped with a choice of structural features, such as a topology or a metric. Then a *spacetime symmetry* is a one-one map from the spacetime to itself that preserves all of this structure. The more structure a spacetime has, the stronger the requirements placed on spacetime symmetries, which means that a spacetime with more structure is one with fewer symmetries.

A dynamical theory specifies a class of nomically possible trajectories on a spacetime that describe the evolution of systems of fields and particles. These trajectories are located on the spacetime, but it can happen that our dynamical theory displays a certain indifference as to how they are located. A *dynamical symmetry* is a one-one map from the spacetime to itself such that a trajectory is nomically possible if and only if its image under the map is also nomically possible. The stronger the dynamical theory, the fewer trajectories it admits as nomically possible, and the fewer dynamical symmetries it possesses. For example, a dynamical symmetry of Newtonian mechanics is translation by a fixed spatial difference. Given a set of trajectories of material objects that form a solution to Newton's laws of motion, we get another solution to the same equations if we shift all objects by the same distance at all times. Newton's laws care about the relative positions of objects, but not about where in space the entire system of objects is located.

John Earman (1989: ch. 3) argues that dynamical symmetries must coincide with spacetime symmetries. He proposes two symmetry principles as adequacy conditions on theories of space, time, and motion:

SP1 Any dynamical symmetry must be a spacetime symmetry
SP2 Any spacetime symmetry must be a dynamical symmetry

The case for SP2 seems fairly straightforward. If we adopt a dynamical theory then we need to endow spacetime with enough structure to support the facts of motion postulated by that theory. If there were a spacetime symmetry that is not a dynamical symmetry then there would be a trajectory that is admitted as nomically possible by our dynamical theory, but whose image under some spacetime symmetry map is prohibited, in spite of the fact that our account of spacetime regards these trajectories as qualitatively indistinguishable. What is at issue here, though, is the converse principle. Earman suggests that SP1 is just an application of Ockham's Razor and I do not think this is correct.

If there were a dynamical symmetry that is not a spacetime symmetry then our theory of spacetime would distinguish trajectories more finely than our dynamics, and thereby postulate differences that play no theoretical role in our best theories of motion. One example where the two types of symmetry come apart is Newton's own account, which combines his own theory of mechanics with substantivalism about both space and time. This account allows us to make sense of the inertial forces at stake in the rotating bucket experiment, but it also permits distinctions that are not needed to account for any observable phenomena. Most notably, it postulates facts about spatial location even though the dynamical theory is invariant under spatial translations. The same is true for my own hybrid view, which inherits the notion of absolute rest and motion from spatial substantivalism, and thus violates SP1.

One can eliminate some of this extra structure by adopting the *neo-Newtonian* spacetime of Lawrence Sklar (1974: 202–6). Instead of one three-dimensional manifold that persists through time, this theory postulates a stack of three-dimensional spaces that succeed one another in time. Because every spatial point only exists for an instant, there is no notion of absolute rest that consists in being located at the same point throughout time. To get a notion of absolute acceleration to account for the observable facts of inertial forces, we equip the spacetime with a bundle of 'inertial trajectories', which are lines that intersect all the three-dimensional spaces and serve as the trajectories of objects that

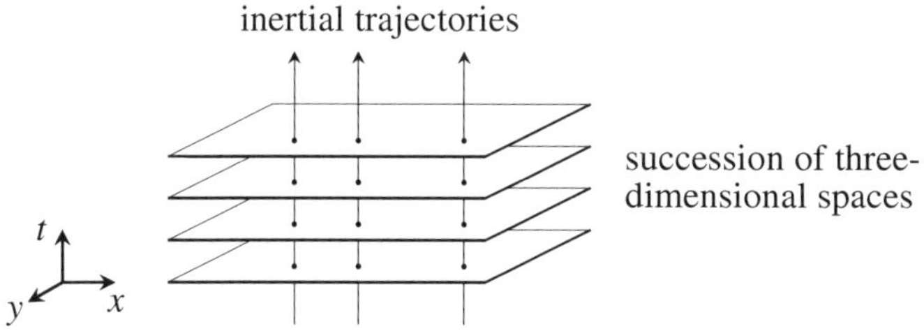

Figure 12.2. Neo-Newtonian spacetime

are not subject to any external forces. This neo-Newtonian spacetime yields absolute acceleration without absolute velocity or location, but it still has *more* structure than is needed for our dynamic theory, and thus still fails to satisfy SP1. If we take all trajectories and shift them by the same fixed distance at all times then we get a different configuration of trajectories in our neo-Newtonian spacetime, but an indistinguishable solution of our theory of motion.

Some applications of SP1 are uncontroversial. We noted earlier that relativity theory only entails the rejection of a frame-independent notion simultaneity if it is supplemented with a parsimony principle that probits us from postulating more spatial and temporal structure than is needed to account for our best physical theories. In this particular instance, SP1 yields the desired result that we cannot simply postulate that some reference frame captures the truth about simultaneity. But there are many cases in which SP1 is impossible to satisfy. Suppose we are given some dynamical theory. If we start out with a structureless four-dimensional spacetime and add the *minimal* amount of structure to satisfy SP2 then there is no guarantee that the resulting account will satisfy SP1, as we just saw in the case of neo-Newtonian spacetime. The problem is that there can be *theoretical bonuses*. Entities that are postulated to address a particular theoretical need, such as spatial points to account for inertial forces, may end up delivering benefits for which there is no such need, such as a notion of absolute location. This is what prevents us from satisfying both SP1 and SP2 in such cases.

In those instances where a structure is detachable from the rest of our account, both Ockham's Razor and SP1 would require that we avoid the extra commitment. But this does not tell us what to do about a theoretical bonus that is provided by structure that already earns its keep in some other way, nor does it help us if there is more than one way of postulating a minimally sufficient amount of structure to satisfy SP2. The neo-Newtonian view might be more frugal than the regular Newtonian view because it does not deliver an absolute notion of velocity, but it is less frugal in that it postulates *more* primitive spacetime points. The interesting question is how to weigh these different types of frugality, and SP1 is simply silent on this issue. Given these concerns about the principle, I do not think that the fact that it violates SP1 counts against my hybrid view of space and time. Our dynamical theories might not need a notion of absolute spatial location, but this does not show that spatial points are superfluous because they already earn their keep by accounting for inertial forces and fields.

13

CONCLUSION

Let me close with a quick overview of my defence of a modal view of time. Some parts of the discussion were concerned with housekeeping tasks that present its various components and explain the inner workings of such a view. Apart from the survey of basic tense logic in Chapter 4, this includes the linguistic ersatzism about times from Chapters 6 and 7, and the quantified tense logic from Chapter 8. The offensive part of the project mainly consists of the dismissal of relationism about time and temporal substantivalism in Chapters 2 and 3. The remainder of the book was devoted to defending modal accounts against the main objections that have been levied against it. There, the key strategy was to separate tense primitivism from other, less plausible theses. Most importantly, this involved rejecting presentism in Chapter 9 and the flow of time in Chapter 10. It also includes the remarks about temporal propositions in Chapter 5, the dismissal of tense logic as a theory of verb tense in the Preface, and the decision to adopt the very weak tense logic Z as a theory of time. Without all of this ballast, it was then fairly easy to reconcile a modal account of time with the theory of relativity in Chapters 11 and 12, and to relegate a host of other questions to different subfields of philosophy, or to empirical inquiry. There are some theses that I took on largely because my commitment to tense primitivism seems to leave me no other choice. This includes the remarks on eternal recurrence in Section 7.2, the account of Gödel's spacetimes without global foliations in Section 12.2, and, to a lesser extent, the decision in Section 7.3 to regard all claims about distances and durations as theory-relative. These might be regarded as costs of the view, but I do not think that they diminish the overall appeal of tense primitivism. The aim was to win the war, not to prevail in every battle.

References

The numerals at the end of bibliographic entries refer to the pages on which the item in question is cited in the main text, or is referred to by other entries in this list of references.

Adams, Robert Merrihew. 1974. Theories of actuality. *Noûs* 8: 211–31. 105

Anderson, Anthony. 1989. Russell on order in time. In Anthony Anderson and Wade Savage, eds., *Rereading Russell*, vol. XII of *Minnesota Studies in the Philosophy of Science*, 249–63. University of Minnesota Press, Minneapolis. 9

Armstrong, D. M. 1996. Dispositions as categorical states. In Tim Crane, ed., *Dispositions: A Debate*, 15–18. Routledge, London. 54

——. 1997. *A World of States of Affairs*. Cambridge University Press, Cambridge. 53

——. 2004. *Truth and Truthmakers*. Cambridge University Press, Cambridge. 54

Arnol'd, Vladimir Igorevich. 1980. *Mathematical Methods of Classical Mechanics*. Springer, New York. 29

Arntzenius, Frank. 2008. Gunk, topology and measure. In Dean Zimmerman, ed., *Oxford Studies in Metaphysics*, vol. 4, 225–47. Clarendon Press, Oxford. 36

Aronszajn, Mark. 1996. A defense of temporalism. *Philosophical Studies* 81: 71–95. 51

Augustine of Hippo. 1999. *Confessions*. Loeb Classical Library. Harvard University Press, Cambridge, Mass. Translated by William Watts. 87

Balashov, Yuri. 2010. *Persistence and Spacetime*. Oxford University Press, Oxford. 124

—— and Janssen, Michel. 2003. Presentism and relativity. *British Journal for the Philosophy of Science* 54: 327–46. 87

Barcan, Ruth. 1946. A functional calculus of first order based on strict implication. *Journal of Symbolic Logic* 11: 1–16. 85

Beebee, Helen and Dodd, Julian, eds. 2005. *Truthmakers: The Contemporary Debate*. Oxford University Press, Oxford. 53

Bell, John. 2004. How to teach special relativity. In *Speakable and Unspeakable in Quantum Mechanics*, 67–80. Cambridge University Press, Cambridge, 2nd ed. 118

Bencivenga, Ermanno. 2002. Free logics. In Dov Gabbay and Franz Guenthner, eds., *Handbook of Philosophical Logic*, vol. 5, 147–96. Kluwer, Dordrecht, 2nd ed. 84

Bennett, Jonathan. 1988. *Events and their Names*. Hackett, Indianapolis. 15, 53

Bennett, Michael. 1977. A guide to the logic of tense and aspect in English. *Logique et Analyse* 20: 491–517. 58

Bergmann, Gustav. 1960. Some reflections on time. In *Meaning and Existence*, 225–63. University of Wisconsin Press, Madison. 20, 25

Bigelow, John. 1988. *The Reality of Numbers: A Physicalist's Philosophy of Mathematics*. Clarendon Press, Oxford. 53

——. 1996. Presentism and properties. In Tomberlin (1996), 35–52. 87, 93, 94

Black, Max. 1959. The 'direction' of time. *Analysis* 19: 54–63. 102

Blackburn, Patrick. 1993. Nominal tense logic. *Notre Dame Journal of Formal Logic* 34: 56–83. 63

Bostock, David. 2010. Whitehead and Russell on points. *Philosophia Mathematica* 18: 1–52. 12

Bourne, Craig. 2002. When am I? A tense time for some tense theorists? *Australasian Journal of Philosophy* 80: 359–71. 91

——. 2006. *A Future for Presentism*. Clarendon Press, Oxford. 66, 87

Braddon-Mitchell, David. 2004. How do we know it is now now? *Analysis* 64: 199–203. 91

Broad, C. D. 1923. *Scientific Thought*. Kegan Paul, London. 87, 100, 104

——. 1933. *Examination of McTaggart's Philosophy*. Cambridge University Press, Cambridge. 102

Brogaard, Berit. 2012. *Transient Truths: An Essay in the Metaphysics of Propositions*. Oxford University Press, New York. 51

Bull, R. A. 1969. On modal logic with propositional quantifiers. *Journal of Symbolic Logic* 34: 257–63. 65

Burgess, John P. 1982. Axioms for tense logic, II: Time periods. *Notre Dame Journal of Formal Logic* 23: 375–83. 58

——. 2002. Basic tense logic. In Gabbay and Guenthner (2002), 1–42. 43, 45

——. 2009. *Philosophical Logic*. Princeton University Press, Princeton. 43

Butterfield, Jeremy. 1985. Spatial and temporal parts. *Philosophical Quarterly* 35: 32–44. 22

Button, Tim. 2006. There's no time like the present. *Analysis* 66: 130–5. 91

——. 2007. Every now and then, no-futurism faces no sceptical problems. *Analysis* 67: 325–32. 91

Callender, Craig. 1998. Shedding light on time. In Howard (1998), 587–99. 90

——, ed. 2002. *Time, Reality & Experience*, vol. 50 of *Royal Institute of Philosophy Supplement*. Cambridge University Press, Cambridge. 153

——. 2011. *The Oxford Handbook of Philosophy of Time*. Oxford University Press, New York. 140, 149

Cappelen, Herman and Hawthorne, John. 2009. *Relativism and Monadic Truth*. Oxford University Press, Oxford. 51

Carathéodory, Constantin. 1956. *Maß und Integral und ihre Algebraisierung*. Birkhäuser, Basel. 36

Carnap, Rudolf. 1925. Über die Abhängigkeit der Eigenschaften des Raumes von denen der Zeit. *Kant-Studien* 30: 331–45. 14

——. 1958. *Introduction to Symbolic Logic and its Applications*. Dover, New York. Translated by William Meyer and John Wilkinson. 21

Carrier, Martin. 2003. How to tell causes from effects: Kant's causal theory of time and modern approaches. *Studies in History and Philosophy of Modern Physics* 34: 59–71. 14

Casati, Roberto and Varzi, Achille, eds. 1996. *Events*. Dartmouth, Aldershot. 15

Chisholm, Roderick. 1976. *Person and Object*. Open Court, Chicago. 14

——. 1990. Referring to things that no longer exist. In James Tomberlin, ed., *Action Theory and Philosophy of Mind*, vol. 4 of *Philosophical Perspectives*, 545–56. Ridgeview, Atascadero. 83

Cocchiarella, Nino. 1966. A completeness theorem for tense logic. *Journal of Symbolic Logic* 31: 689–91. 45

——. 1969. Existence entailing attributes, modes of copulation and modes of being in second order logic. *Noûs* 3: 33–48. 83

——. 2007. *Formal Ontology and Conceptual Realism.* Springer, Dordrecht. 81, 82

Cresswell, M. J. 1970. Classical intensional logic. *Theoria* 36: 347–72. 42

——. 1990. *Entities and Indices.* Kluwer, Dordrecht. 58

——. 1996. *Semantic Indexicality.* Kluwer, Dordrecht. 26

——. 2010. Temporal reference in linear tense logic. *Journal of Philosophical Logic* 39: 173–200. 63

Crisp, Thomas. 2003. Presentism. In Michael Loux and Dean Zimmerman, eds., *The Oxford Handbook of Metaphysics*, 211–45. Oxford University Press, New York. 90

——. 2004. On presentism and triviality. In Zimmerman (2004), 15–46. Reprinted in Magalhães and Oaklander (2010). 87

——. 2007. Presentism and the grounding objection. *Noûs* 41: 118–37. 93, 94

Crossley, John and Humberstone, Lloyd. 1977. The logic of 'actually'. *Reports on Mathematical Logic* 8: 11–29. 58

Dainton, Barry. 2008. The experience of time and change. *Philosophy Compass* 3: 619–38. 102

——. 2011. Time, passage, and immediate experience. In Callender (2011), 381–418. 102

Davey, B. A. and Priestley, H. A. 1990. *Introduction to Lattices and Order.* Cambridge University Press, Cambridge. 13

Davidson, Donald. 1980a. Causal relations. In Davidson (1980b), 149–62. 15

——. 1980b. *Essays on Actions and Events.* Clarendon Press, Oxford. 140

——. 1980c. The logical form of action sentences. In Davidson (1980b), 105–22. 15, 53

Dodd, Julian. 2002. Is truth supervenient on being? *Proceedings of the Aristotelian Society* 102: 69–85. 53

Dorato, Mauro. 2006. The irrelevance of the presentist/eternalist debate for the ontology of Minkowski spacetime. In Dennis Dieks, ed.,

The Ontology of Spacetime II, 93–109. Elsevier, Amsterdam and New York. 90

Dretske, Fred. 1961. Particulars and the relational theory of time. *Philosophical Review* 70: 447–69. 20

Dummett, Michael. 1960. A defense of McTaggart's proof of the unreality of time. *Philosophical Review* 69: 497–504. 110

——. 1981. *Frege: Philosophy of Language*. Harvard University Press, Cambridge, Mass., 2nd ed. 24

——. 2000. Is time a continuum of instants? *Philosophy* 75: 497–515. 36

——. 2003. Truth and the past. *Journal of Philosophy* 100: 5–53. 92

——. 2005. Hume's atomism about events: a response to Ulrich Meyer. *Philosophy* 80: 141–4. 36

Earman, John. 1970. Space-time, or how to solve philosophical problems and dissolve philosophical muddles without really trying. *Journal of Philosophy* 67: 259–77. 68, 123

——. 1989. *World Enough and Spacetime*. The MIT Press, Cambridge, Mass. 31, 132

——. 1995. *Bangs, Crunches, Whimpers, and Shrieks*. Oxford University Press, New York. 69

——, Glymour, Clark, and Stachel, John, eds. 1977. *Foundations of Space-Time Theories*, vol. VIII of *Minnesota Studies in the Philosophy of Science*. University of Minnesota Press, Minneapolis. 154

Eddon, Maya. 2010. Three arguments from temporary intrinsics. *Philosophy and Phenomenological Research* 81: 605–19. 26

Einstein, Albert. 1918. *Über die spezielle und die allgemeine Relativitätstheorie*. Vieweg, Braunschweig, 3rd ed. 111, 116, 118

——. 1921. *Geometrie und Erfahrung*. Springer, Berlin. 119

——. 1929. Spacetime. In *Enyclopædia Britannica*, vol. 21, 105–8. The Enyclopædia Britannica Company, London, 14th ed. 119

——. 1956. *The Meaning of Relativity*. Princeton University Press, Princeton, 5th ed. 118, 119, 123

Euclid. 1883–1916. *Opera omnia*. Teubner, Leipzig. Edited by I. L. Heiberg and H. Menge. 111

Euler, Leonhard. 1748. Réflexions sur l'espace et le temps. *Mémoires de l'academie des sciences de Berlin* 4: 324–33. Reprinted in *Leonhardi Euleri Opera Omnia*, ser. 3, vol. II, Teubner, Leipzig, 1942. 72

Field, Hartry. 1980. *Science without Numbers.* Princeton University Press, Princeton. 32

——. 1984. Can we dispense with spacetime? In Peter Asquith and Philip Kitcher, eds., *PSA 1984: Proceedings of the 1984 Biennial Meeting of the Philosophy of Science Association*, vol. 2, 33–90. Philosophy of Science Association, East Lansing. 32

Fine, Kit. 1970. Propositional quantifiers in modal logic. *Theoria* 36: 336–46. 65

——. 2005a. *Modality and Tense.* Clarendon Press, Oxford. 142

——. 2005b. Prior on the construction of possible worlds and instants. In Fine (2005a), 133–75. 65

——. 2005c. Tense and reality. In Fine (2005a), 261–320. 53

Fiocco, Oreste. 2007. Passage, becoming and the nature of temporal reality. *Philosophia* 35: 1–21. 100

Fitch, G. W. 1998. Temporalism revisited. *Philosophical Studies* 92: 251–6. 51

Flo, Olav. 1970. Bibliography of the philosophical writings of A. N. Prior. *Theoria* 36: 189–213. vii

Forbes, Graeme. 1993. Time, events, and modality. In Robin Le Poidevin and Murray MacBeath, eds., *The Philosophy of Time*, Oxford Readings in Philosophy, 80–95. Oxford University Press, New York. 11

Forrest, Peter. 2004. The real but dead past: a reply to Braddon-Mitchell. *Analysis* 64: 358–62. 91

Friedman, Michael. 2002. Geometry as a branch of physics: Background and context for Einstein's 'Geometry and Experience'. In Malament (2002), 193–229. 119

Gabbay, Dov. 1975. Model theory for tense logics. *Annals of Mathematical Logic* 8: 185–236. 45

——. 1977. A tense system with split truth table. *Logique et Analyse* 20: 359–93. 57

—— and Guenthner, Franz, eds. 2002. *Handbook of Philosophical Logic*, vol. 7. Kluwer, Dordrecht, 2nd ed. 139, 146

—— and Malod, Guillaume. 2002. Naming worlds in modal and temporal logic. *Journal of Logic, Language and Information* 11: 29–65. 63

Gale, Richard. 1968. *The Language of Time.* Routledge & Kegan Paul, London. 100

——. 1969. ‘Here’ and ‘now’. *The Monist* 53: 396–409. 100

——. 1996. Disanalogies between space and time. *Process-Studies* 25: 72–89. 100

Garrett, Brian. 1988. ‘Thank goodness that’s over’ revisited. *Philosophical Quarterly* 38: 201–5. 52

Garson, James. 1969. Here and now. *The Monist* 53: 469–77. 100

——. 2001. Quantification in modal logic. In Dov Gabbay and Franz Guenthner, eds., *Handbook of Philosophical Logic*, vol. 3, 267–323. Kluwer, Dordrecht, 2nd ed. 83

Giulini, Domenico. 2005. *Special Relativity: A First Encounter*. Oxford University Press, Oxford. 116

Gödel, Kurt. 1949a. An example of a new type of cosmological solutions of Einstein’s field equations of gravitation. *Review of Modern Physics* 21: 447–50. 123

——. 1949b. A remark about the relationship between relativity theory and idealistic philosophy. In Paul Schilpp, ed., *Albert Einstein: Philosopher–Scientist*, 557–62. Library of Living Philosophers, Evanston. 123

Goldstein, Herbert. 1980. *Classical Mechanics*. Addison–Wesley, Boston, 2nd ed. 29

Goodman, Nelson. 1955. *Fact, Fiction, and Forecast*. Harvard University Press, Cambridge, Mass. 56

Grünbaum, Adolf. 1953. Whitehead’s method of extensive abstraction. *British Journal for the Philosophy of Science* 4: 215–26. 12

——. 1955. Logical and philosophical foundations of the special theory of relativity. *American Journal of Physics* 23: 450–64. 119

——. 1962. The relevance of philosophy to the history of the special theory of relativity. *Journal of Philosophy* 59: 561–73. 119

——. 1963a. Carnap’s views on the foundations of geometry. In Paul A. Schilpp, ed., *The Philosophy of Rudolf Carnap*, 599–684. Open Court, Chicago. 14

——. 1963b. *Philosophical Problems of Space and Time*. Knopf, New York. 14

——. 1967. *Modern Science and Zeno’s Paradoxes*. Wesleyan University Press, Middletown. 14

——. 1968. *Geometry and Chronometry in Philosophical Perspective*. University of Minnesota Press, Minneapolis. 72

Grzegorczyk, Andrzej. 1960. Axiomatizability of geometry without points. *Synthese* 12: 228–35. 36

Hacking, Ian. 1994. What is logic? In Dov M. Gabbay, ed., *What is a Logical System?*, vol. 4 of *Studies in Logic and Computation*, 1–33. Oxford University Press, New York. 47

Hamblin, C. L. 1970. The effect of when it's said. *Theoria* 36: 249–63. 51

Hammerschmidt, William. 1947. *Whitehead's Philosophy of Time.* King's Crown Press, New York. 8

Haslanger, Sally. 1989a. Endurance and temporary intrinsics. *Analysis* 49: 119–25. 4

——. 1989b. Persistence, change, and explanation. *Philosophical Studies* 56: 1–28. 4

Hawley, Katherine. 1998. Why temporary properties are not relations between physical objects and times. *Proceedings of the Aristotelian Society* 98: 211–16. 26

——. 2001. *How Things Persist.* Clarendon Press, Oxford. 21

Hazen, Allen. 1976. Expressive completeness in modal languages. *Journal of Philosophical Logic* 5: 25–46. 58

——. 1978. The eliminability of the actuality operator in propositional modal logic. *Notre Dame Journal of Formal Logic* 19: 617–22. 57

Heller, Mark. 1990. *The Ontology of Physical Objects.* Cambridge University Press, Cambridge. 21

——. 1992. Things change. *Philosophy and Phenomenological Research* 52: 695–704. 21

Hempel, Carl. 1958. The theoretician's dilemma. In Herbert Feigl, Michael Scriven, and Gover Maxwell, eds., *Concepts, Theories, and the Mind-Body Problem*, vol. II of *Minnesota Studies in the Philosophy of Science*, 37–98. University of Minnesota Press, Minneapolis. 74

——. 1966. *Philosophy of Natural Science.* Prentice Hall, Englewood Cliffs. 74

Hestevold, Scott and Carter, William. 2002. On presentism, endurance, and change. *Canadian Journal of Philosophy* 32: 491–510. 94

Higginbotham, James. 2003. Tensed second thoughts. In Jokić and Smith (2003), 191–7. 52

——, Pianesi, Fabio, and Varzi, Achille, eds. 2000. *Speaking of Events.* Oxford University Press, New York. 15

Hinchliff, Mark. 1996. The puzzle of change. In Tomberlin (1996), 119–36. 4

Hobbes, Thomas. 1651. *Leviathan*. Andrew Crooke, London. 87, 101

Hodes, Harold. 1984. Some theorems on the expressive limitations of modal languages. *Journal of Philosophical Logic* 13: 13–26. 58

Hoerl, Christoph. 2009. Time and tense in perceptual experience. *Philosophers' Imprint* 9. 102

Howard, Don, ed. 1998. *PSA 1998: Proceedings of the 1998 Biennial Meeting of the Philosophy of Science Association*. Philosophy of Science Association, East Lansing. 139, 153

Hughes, G. E. and Cresswell, M. J. 1975. Omnitemporal logic and converging time. *Theoria* 41: 11–34. 58

——. 1996. *A New Introduction to Modal Logic*. Routledge, London. 42, 83, 85

Humberstone, Lloyd. 1978. Interval semantics for tense logic: Some remarks. *Journal of Philosophical Logic* 8: 171–96. 58

James, William. 1890. *The Principles of Psychology*. Holt, New York. 101

Jammer, Max. 1954. *Concepts of Space*. Harvard University Press, Cambridge, Mass. 119

Johnstone, P. T. 1982. *Stone Spaces*. Cambridge University Press, Cambridge. 13

Jokić, Aleksandar and Smith, Quentin, eds. 2003. *Time, Tense, and Reference*. The MIT Press, Cambridge, Mass. 144, 152

Kamp, Hans. 1968. *Tense Logic and the Theory of Linear Orders*. Ph.D. thesis, University of California, Los Angeles. 57

——. 1971. Formal properties of 'now'. *Theoria* 37: 227–73. 57

——. 1979. Events, instants and temporal reference. In Rainer Bäuerle, Urs Egli, and Armin von Stechow, eds., *Semantics from Different Points of View*, 376–417. De Gruyter, Berlin. 9, 12, 58

Kaplan, David. 1970. S5 with quantifiable propositional variables. *Journal of Symbolic Logic* 35: 355. 65

Keller, Simon. 2004. Presentism and truthmaking. In Zimmerman (2004), 83–104. 93

Kim, Jaegwon. 1993. Events as property exemplifications. In *Supervenience and Mind*, 33–52. Cambridge University Press, Cambridge. 14

Kripke, Saul. 1980. *Naming and Necessity*. Blackwell, Oxford. 19, 82, 93

Kuhn, Thomas. 2002. Tense and time. In Gabbay and Guenthner (2002), 277–346. 58

Lacey, Hugh. 1968. The causal theory of time: A critique of Grünbaum's version. *Philosophy of Science* 35: 332–54. 14

de Laguna, Theodore. 1922a. The nature of space. *Journal of Philosophy* 19: 393–407 and 421–40. 12

——. 1922b. Point, line, and surface, as sets of solids. *Journal of Philosophy* 19: 449–61. 12

Le Poidevin, Robin. 1991. *Change, Cause and Contradiction: A Defence of the Tenseless Theory of Time*. St. Martin's Press, New York. 110

——. 2003. *Travels in Four Dimensions: The Enigmas of Space and Time*. Clarendon Press, Oxford. 14

——. 2007. *The Images of Time*. Oxford University Press, New York. 100

Leibniz, Gottfried Wilhelm. 1863. Initia rerum mathematicarum metaphysica. In *Leibnizens mathematische Schriften*, 17–29. H. W. Schmidt, Halle. Edited by Carl Immanuel Gerhardt. 14

——. 1875–90. *Die philosophischen Schriften von Gottfried Wilhelm Leibniz*. Weidmannsche Buchhandlung, Berlin. Edited by Carl Immanuel Gerhardt. Reprinted by Georg Olms, Hildesheim, 1996. 28

Lemmon, E. J. 1966. Comments on Donald Davidson, 'The logical form of action sentences'. In Nicholas Rescher, ed., *The Logic of Decision and Action*, 96–103. University of Pittsburgh Press, Pittsburgh. 14

Levison, Arnold. 1987. Events and time's flow. *Mind* 96: 341–53. 100

Lewis, David. 1973. *Counterfactuals*. Harvard University Press, Cambridge, Mass. 56

——. 1983a. How to define theoretical terms. In Lewis (1983b), 78–95. 74

——. 1983b. *Philosophical Papers*, vol. I. Oxford University Press, New York. 21, 146

——. 1986a. Causation. In Lewis (1986d), 159–213. 15

——. 1986b. Events. In Lewis (1986d), 241–69. 14, 15

——. 1986c. *On the Plurality of Worlds*. Blackwell, Oxford. 4, 20, 26, 59, 93, 96

——. 1986d. *Philosophical Papers*, vol. II. Oxford University Press, New York. 146

——. 1991. *Parts of Classes*. Blackwell, Oxford. 91

——. 1992. Critical notice of D. M. Armstrong, *A Combinatorial Theory of Possibility*. *Australasian Journal of Philosophy* 70: 211–24. 53, 55

——. 2004. Tensed quantifiers. In Zimmerman (2004), 3–14. 81

Linsky, Bernard and Zalta, Edward. 1994. In defense of the simplest quantified modal logic. In James Tomberlin, ed., *Logic and Language*, vol. 8 of *Philosophical Perspectives*, 431–58. Ridgeview, Atascadero. 83

Lombard, Lawrence. 1986. *Events: A Metaphysical Study*. Routledge & Kegan Paul, London. 15

——. 1994. The doctrine of temporal parts and the 'no-change' objection. *Philosophy and Phenomenological Research* 54: 365–72. 21

——. 1999. On the alleged incompatibility of presentism and temporal parts. *Philosophia* 27: 253–60. 90

Lorentz, Hendrik. 1913. *Das Relativitätsprinzip*. Teubner, Leipzig. 118

Lowe, E. J. 1987. The indexical fallacy in McTaggart's proof of the unreality of time. *Mind* 96: 62–70. 110

——. 1992. McTaggart's paradox revisited. *Mind* 101: 323–6. 110

——. 1998. *The Possibility of Metaphysics: Substance, Identity, and Time*. Clarendon Press, Oxford. 110

Lucretius, Titus. 1947. *De rerum natura*. Clarendon Press, Oxford. Edited and translated by Cyril Bailey. 87, 93

Ludlow, Peter. 2004. Presentism, triviality, and the varieties of tensism. In Zimmerman (2004), 21–36. 87

MacBeath, Murray. 1983. Mellor's emeritus headache. *Ratio* 25: 81–8. 52

Mach, Ernst. 1912. *Die Mechanik in ihrer Entwicklung*. Brockhaus, Leipzig, 7th ed. 73

Magalhães, Ernâni and Oaklander, Nathan, eds. 2010. *Presentism: Essential Readings*. Rowman & Littlefield, Lanham. 140, 148

Malament, David, ed. 2002. *Reading Natural Philosophy*. Open Court, Chicago. 142

Marcus, Ruth Barcan. 1993. *Modalities: Philosophical Essays*. Oxford University Press, New York. 78

Markosian, Ned. 1993. How fast does time pass? *Philosophy and Phenomenological Research* 53: 829–44. 102

——. 2004. A defense of presentism. In Zimmerman (2004), 47–82. 87

Maudlin, Tim. 2002. Remarks on the passing of time. *Proceedings of the Aristotelian Society* 102: 237–52. 102

McArthur, Robert. 1976. *Tense Logic*, vol. 111 of *Synthese Library*. Reidel, Dordrecht. 43

McCall, Storrs. 1976. Objective time flow. *Philosophy of Science* 43: 337–62. 100

McKitrick, Jennifer. 2003. The bare metaphysical possibility of bare dispositions. *Philosophy and Phenomenological Research* 66: 349–69. 54

McTaggart, J. M. E. 1927. *The Nature of Existence*. Cambridge University Press, Cambridge. 100, 109

Mead, George. 1932. *The Philosophy of the Present*. Open Court, Chicago. 87

Mellor, D. H. 1974. In defense of dispositions. *Philosophical Review* 83: 157–81. 54

——. 1981a. *Real Time*. Cambridge University Press, Cambridge. 25, 26, 110

——. 1981b. 'Thank goodness that's over'. *Ratio* 23: 20–30. 52

——. 1983. Macbeath's soluble aspirin. *Ratio* 25: 89–92. 52

——. 1998. *Real Time II*. Routledge, London. 21, 26, 53, 110, 131

Merricks, Trenton. 2006. Goodbye Growing Block. In Dean Zimmerman, ed., *Oxford Studies in Metaphysics*, vol. 2, 103–10. Clarendon Press, Oxford. 91

——. 2007. *Truth and Ontology*. Clarendon Press, Oxford. 53, 87

——. 2009. Propositional attitudes? *Proceedings of the Aristotelian Society* 109: 207–32. 52

Meyer, Ulrich. 2002. Is science first-order? *Analysis* 62: 305–8. 54

——. 2005a. Dummett on the time-continuum. *Philosophy* 80: 135–40. 36

——. 2005b. The presentist's dilemma. *Philosophical Studies* 122: 213–25. Reprinted in Magalhães and Oaklander (2010). 90

——. 2006. Worlds and times. *Notre Dame Journal of Formal Logic* 47: 25–37. 98

——. 2009a. 'Now' and 'then' in tense logic. *Journal of Philosophical Logic* 38: 229–47. 57, 82

——. 2009b. Times in tense logic. *Notre Dame Journal of Formal Logic* 50: 201–19. 57, 61, 63, 68

——. 2011. Time and modality. In Callender (2011), 91–121. vii

Minkowski, Hermann. 1909. Raum und Zeit. *Physikalische Zeitschrift* 10: 104–11. 119

Montague, Richard. 1969. On the nature of certain philosophical entities. *The Monist* 53: 159–94. 16

Mormann, Thomas. 1998. Continuous lattices and Whiteheadian theory of space. *Logic and Logical Philosophy* 6: 35–54. 13

——. 2005. Description, construction, and representation: From Russell and Carnap to Stone. In Bernard Linsky and Guido Imaguire, eds., *On Denoting: 1905–2005*, 333–60. Philosophia, Munich. 13

Mulligan, Kevin, Simons, Peter, and Smith, Barry. 1984. Truth-makers. *Philosophy and Phenomenological Research* 44: 287–321. 55

Myro, George. 1986a. Existence and time. In Richard Grandy and Richard Warner, eds., *The Philosophical Grounds of Rationality*, 383–409. Clarendon Press, Oxford. 2, 60

——. 1986b. Time and essence. In Peter French, Theodore Uehling, and Howard Wettstein, eds., *Studies in Essentialism*, 331–41, vol. 11 of *Midwest Studies in Philosophy*. University of Minnesota Press, Minneapolis. 2, 60

Newton, Isaac. 1687. *Philosophia naturalis principia mathematica*. Joseph Streater, London. 29

Newton-Smith, W. H. 1980. *The Structure of Time*. Routledge, London. 71

Noether, Emmy. 1918. Invariante Variationsprobleme. *Nachrichten von der Königlichen Gesellschaft der Wissenschaften zu Göttingen, Mathematisch–physikalische Klasse*, 235–57. 29

Oaklander, L. Nathan. 1983. McTaggart, Schlesinger and the two-dimensional time hypothesis. *Philosophical Quarterly* 33: 391–7. 102

——. 1984. *Temporal Relations and Temporal Becoming: A Defense of a Russellian Theory of Time*. University Press of America, Lanham. 102

——. 1985. A reply to Schlesinger. *Philosophical Quarterly* 35: 93–4. 102

Oaklander, L. Nathan and Smith, Quentin, eds. 1994. *The New Theory of Time*. Yale University Press, New Haven. 110, 154

Olson, Eric. 2009. The rate of time's passage. *Analysis* 69: 3–9. 102

Ørstrøm, Peter and Hasle, Per. 1993. A. N. Prior's rediscovery of tense logic. *Erkenntnis* 39: 23–50. vii

Pap, Arthur. 1962. *An Introduction to the Philosophy of Science*. Free Press of Glencoe, New York. 71

Parsons, Terence. 1967. Grades of essentialism in quantified modal logic. *Noûs* 1: 181–91. 78

——. 1969. Essentialism and quantified modal logic. *Philosophical Review* 78: 35–52. 78

Paul, L. A. 2010. Temporal experience. *Journal of Philosophy* 107: 333-59. 102

Perry, John. 2001. Time, consciousness and the knowledge argument. In Nathan Oaklander, ed., *The Importance of Time*, 81–93. Kluwer, Dordrecht. 51

Phillips, Ian. 2009. Rate abuse: A reply to Olson. *Analysis* 69: 503–5. 102

Pianesi, Fabio and Varzi, Achille. 1996. Events, topology, and temporal relations. *The Monist* 79: 89–116. 11

Pnueli, Amir. 1977. The temporal logic of programs. In *18th Annual Symposium of Computer Science*, 46–57. Institute of Electrical and Electronics Engineers, New York. 43

Poincaré, Henri. 1902. La science et l'hypothèse. Flammarion, Paris. 75

——. 1906. Sur la dynamique de l'électron. *Rendiconti del Circolo Matematico di Palermo* 21: 129–75. 119

Pooley, Oliver. 2013. Substantivalist and relationalist approaches to spacetime. In Robert Batterman, ed., *The Oxford Handbook of Philosophy of Physics*, 522–86. Oxford University Press, New York. 31

Price, Huw. 1996. *Time's Arrow and Archimedes' Point*. Oxford University Press, New York. 102

Prior, Arthur. 1957a. A tense-logical analogue of S5. In Prior (1957b), 18–28. 43

——. 1957b. *Time and Modality*. Clarendon Press, Oxford. vii, 57, 150

——. 1959. Thank goodness that's over. *Philosophy* 34: 12–17. 52

——. 1962. The formalities of omniscience. *Philosophy* 37: 114–29. 51

——. 1966. Postulates for tense logic. *American Philosophical Quarterly* 3: 153–61. 45

——. 1967. *Past, Present and Future*. Clarendon Press, Oxford. vii, 65, 81, 83

——. 1968a. 'Now'. *Noûs* 2: 101–19. 57

——. 1968b. 'Now' corrected and condensed. *Noûs* 2: 411–12. 57

——. 1968c. *Papers on Time and Tense*. Clarendon Press, Oxford. vii

——. 1972. The notion of the present. In J. T. Fraser, F. C. Haber, and G. H. Müller, eds., *The Study of Time*, 320–3. Springer, Berlin. 87

Prosser, Simon. 2012. Why does time seem to pass? *Philosophy and Phenomenological Research* 85: 92-116. 102

Putnam, Hilary. 1967. Time and physical geometry. *Journal of Philosophy* 64: 240–7. 87, 124

——. 2008. Reply to Mauro Dorato. *European Journal of Analytic Philosophy* 4: 71–3. 87

Quine, W. V. 1951. Two dogmas of empiricism. *Philosophical Review* 60: 20–43. 47

——. 1953. Three grades of modal involvement. In *The Ways of Paradox and Other Essays*, 158–76. Harvard University Press, Cambridge, Mass., 2nd ed. 78

——. 1960. *Word and Object*. The MIT Press, Cambridge, Mass. 14, 38

——. 1969. Propositional objects. In *Ontological Relativity and Other Essays*, 139–60. Harvard University Press, Cambridge, Mass. 67

Reichenbach, Hans. 1924a. *Axiomatik der relativistischen Raum-Zeit-Lehre*. Vieweg, Braunschweig. An English translation by Maria Reichenbach, *Axiomatization of the Theory of Relativity*, was published by University of California Press in 1969. 14

——. 1924b. Die Bewegungslehre bei Newton, Leibniz und Huygens. *Kant-Studien* 29: 416–38. 14

——. 1928. *Philosophie der Raum-Zeit-Lehre*. De Gruyter, Berlin. An English translation by Maria Reichenbach, *The Philosophy of Space & Time*, was published by Dover in 1957. 14

——. 1951. *The Rise of Scientific Philosophy*. University of California Press, Berkeley and Los Angeles. 14

Reichenbach, Hans. 1956. *The Direction of Time*. University of California Press, Berkeley and Los Angeles. 14

Rescher, Nicholas. 1966. On the logic of chronological propositions. *Mind* 75: 75–96. 50, 60, 81

—— and Garson, James. 1968. Topological logic. *Journal of Symbolic Logic* 33: 537–48. 26

Reynolds, Mark. 1994. Axiomatisation and decidability of F and P in cyclical time. *Journal of Philosophical Logic* 23: 197–224. 71

Richard, Mark. 1981. Temporalism and eternalism. *Philosophical Studies* 39: 1–13. 51

——. 2003. Objects of relief. In Jokić and Smith (2003), 157–89. 52

Rietdijk, C. W. 1966. A rigorous proof of determinism derived from the special theory of relativity. *Philosophy of Science* 33: 341–4. 87

Robb, Alfred. 1914. *A Theory of Time and Space*. Cambridge University Press, Cambridge. 121

——. 1921. *The Absolute Relations of Time and Space*. Cambridge University Press, Cambridge. 121

Rodriguez-Pereyra, Gonzalo. 2006. Truthmakers. *Philosophy Compass* 1: 186–200. 53

Roeper, Peter. 1997. Region-based topology. *Journal of Philosophical Logic* 26: 251–309. 13

Roper, Andrew. 1982. Towards an eliminative reduction of possible worlds. *Philosophical Quarterly* 32: 45–59. 59

Rosen, Gideon. 1992. *Remarks on Modern Nominalism*. Ph.D. thesis, Princeton University. 18

Russell, Bertrand. 1901. Is position in time and space absolute or relative? *Mind* 10: 293–317. 15

——. 1915. *Our Knowledge of the External World as a Field for Scientific Method in Philosophy*. Open Court, Chicago. 9, 10

——. 1927. *The Analysis of Matter*. Kegan Paul, London. 9

——. 1936. On order in time. *Proceedings of the Cambridge Philosophical Society* 32: 216–28. 9

——. 1948. *Human Knowledge: Its Scope and Limits*. Simon and Schuster, New York. 9

——. 1969. *The ABC of Relativity*. Allen & Unwin, London, 3rd ed. 123

Ryle, Gilbert. 1949. *The Concept of Mind*. Hutchinson, London. 54

Salmon, Nathan. 1986. *Frege's Puzzle*. The MIT Press, Cambridge, Mass. 51

——. 1989. Tense and singular propositions. In Joseph Almog, John Perry, and Howard Wettstein, eds., *Themes from Kaplan*, 331–92. Oxford University Press, New York. 51

Sanson, David and Caplan, Ben. 2010. The way things were. *Philosophy and Phenomenological Research* 81: 24–39. 93

Sattig, Thomas. 2006. *The Language and Reality of Time*. Clarendon Press, Oxford. 124

Saunders, Simon. 2002. How relativity contradicts presentism. In Callender (2002), 277–92. 87

Savitt, Steven. 1998. There's no time like the present (in Minkowski spacetime). In Howard (1998), 563–74. 87

——. 2001. A limited defense of passage. *American Philosophical Quarterly* 38: 261–70. 102

——. 2002. On absolute becoming and the myth of passage. In Callender (2002), 153–67. 102, 103

——. 2005. Time travel and becoming. *The Monist* 88: 413–22. 102, 123

——. 2006. Presentism and eternalism in perspective. In Dennis Dieks, ed. *The Ontology of Spacetime I*, 111–27. Elsevier, Amsterdam and New York. 90

Schlesinger, George. 1970. Change and time. *Journal of Philosophy* 67: 294–300. 68

——. 1980. *Aspects of Time*. Hackett, Indianapolis. 102, 104

——. 1985. How to navigate the river of time. *Philosophical Quarterly* 35: 90–2. 102

——. 1991. E pur si muove. *Philosophical Quarterly* 41: 427–41. 100, 102

——. 1994. *Timely Topics*. St. Martin's Press, New York. 102

Schlick, Moritz. 1915. Die philosophische Bedeutung des Relativitätsprinzips. *Zeitschrift für Philosophie und philosophische Kritik* 159: 129–75. 118

Schopenhauer, Arthur. 1859. *Die Welt als Wille und Vorstellung*. Hendel, Halle, 3rd ed. 87

Segerberg, Krister. 1973. Two-dimensional modal logic. *Journal of Philosophical Logic* 2: 77–97. 57

Shoemaker, Sydney. 1969. Time without change. *Journal of Philosophy* 66: 363–81. 68

——. 1980. Causality and properties. In Peter van Inwagen, ed., *Time and Cause*, 109–35. Reidel, Dordrecht. 107

Sider, Theodore. 1999. Presentism and ontological commitment. *Journal of Philosophy* 96: 325–47. 87, 94

——. 2001. *Four-Dimensionalism*. Clarendon Press, Oxford. 20, 21, 87, 124

Simons, Peter. 1987. *Parts: A Study in Ontology*. Clarendon Press, Oxford. 21

——. 2006. The logic of location. *Synthese* 150: 443–58. 26

Sklar, Lawrence. 1974. *Space, Time, and Spacetime*. University of California Press, Berkeley and Los Angeles. 14, 132

——. 1981. Time, reality, and relativity. In Richard Healy, ed., *Reduction, Time, and Reality*, 129–42. Cambridge University Press, Cambridge. 87

Skow, Bradford. 2007. What makes time different from space? *Noûs* 41: 227–52. 125

Smart, J. J. C. 1949. The river of time. *Mind* 58: 483–94. 102, 110

——. 1972. Space-time and individuals. In Richard Rudner and Israel Scheffler, eds., *Logic & Art: Essays in Honor of Nelson Goodman*, 3–20. Bobbs-Merrill, Indianapolis. 124

Smith, Quentin. 1994. General introduction: The implications of the tensed and tenseless theories of time. In Oaklander and Smith (1994), 1–14. 55

Stalnaker, Robert. 1968. A theory of conditionals. In Nicholas Rescher, ed., *Studies in Logical Theory*, vol. 2 of *American Philosophical Quarterly Monograph Series*, 98–112. Blackwell, Oxford. 56

——. 1976. Possible worlds. *Noûs* 10: 65–75. 99

Stein, Howard. 1968. On Einstein–Minkowski space-time. *Journal of Philosophy* 65: 5–23. 87, 100

——. 1970. On the paradoxical time-structures of Gödel. *Philosophy of Science* 37: 589–601. 123

——. 1977. Some philosophical prehistory of general relativity. In Earman et al. (1977), 3–49. 119

——. 1991. On relativity theory and openness of the future. *Philosophy of Science* 58: 147–67. 100

Strawson, Peter. 1959. *Individuals*. Methuen, London. 14, 20

Tallant, Jonathan. 2007. There have been, are (now), and will be lots of times like the present in the hybrid view of time. *Analysis* 67: 83–6. 91

——. 2011. There's no future in no-futurism. *Erkenntnis* 74: 37–52. 91

Taylor, J. G. 1975. *Special Relativity*. Clarendon Press, Oxford. 116

Taylor, Richard. 1960. Pure becoming. *Australasian Journal of Philosophy* 38: 137–43. 106

——. 1974. *Metaphysics*. Prentice–Hall, Englewood Cliffs, 2nd ed. 106

Thomason, S. K. 1974. Reduction of tense logic to modal logic, I. *Journal of Symbolic Logic* 39: 549–51. 58

——. 1975. Reduction of tense logic to modal logic, II. *Theoria* 41: 154–69. 58

——. 1984. On constructing instants from events. *Journal of Philosophical Logic* 13: 85–96. 9, 11

——. 1989. Free construction of time from events. *Journal of Philosophical Logic* 18: 43–67. 9

Thomson, Judith Jarvis. 1983. Parthood and identity across time. *Journal of Philosophy* 80: 201–20. 21

Tomberlin, James, ed. 1996. *Metaphysics*, vol. 10 of *Philosophical Perspectives*. Blackwell, Cambridge, Mass. 138, 145

Tooley, Michael. 1972. Armstrong's proof of the realist account of dispositional properties. *Australasian Journal of Philosophy* 50: 283–7. 54

——. 1997. *Time, Tense, and Causation*. Clarendon Press, Oxford. 14, 87, 125

van Benthem, Johan. 1977. Tense logic and standard logic. *Logique et Analyse* 80: 395–437. 58

——. 1980. Points and periods. In Christian Rohrer, ed., *Time, Tense and Quantifiers*, 39–57. Niemeyer, Tübingen. 58

——. 1983. *The Logic of Time*. Reidel, Dordrecht. 14, 43

van Fraassen, Bas. 1970. *An Introduction to the Philosophy of Time and Space*. Random House, New York. 14, 27, 71, 118

Vlach, Frank. 1973. *'Now' and 'Then': A Formal Study in the Logic of Tense Anaphora*. Ph.D. thesis, University of California, Los Angeles. 58

von Wright, Georg Henrik. 1968. *Time, Change, and Contradiction*. Cambridge University Press, Cambridge. 5

——. 1979. A modal logic of place. In Ernest Sosa, ed. *The Philosophy of Nicholas Rescher*, 65–73. Reidel, Dordrecht. 26

Walker, A. G. 1947. Durées et instants. *Revue Scientifique* 85: 131–4. 9, 10, 11

Warmbrōd, Ken. 2005. Temporal vacua. *Philosophical Quarterly* 54: 266–86. 68

Wasserman, Ryan. 2003. The argument from temporary intrinsics. *Australasian Journal of Philosophy* 81: 413–19. 26

Webb, C. W. 1960. Could time flow? If so, how fast? *Journal of Philosophy* 57: 357–65. 102

Weingard, Robert. 1972. Relativity and the reality of past and future events. *British Journal for the Philosophy of Science* 23: 119–21. 87

Whitehead, Alfred North. 1919. Time, space, and material. *Aristotelian Society Supplement* 2: 44–57. 8

——. 1920. *The Concept of Nature*. Cambridge University Press, Cambridge. 8

——. 1925. *An Enquiry Concerning the Principles of Natural Knowledge*. Cambridge University Press, Cambridge, 2nd ed. 8, 17

——. 1929. *Process and Reality*. Cambridge University Press, Cambridge. 8, 12

——. 1938. *Modes of Thought*. Macmillan, New York. 8

Whitrow, G. J. 1980. *The Natural Philosophy of Time*. Clarendon Press, Oxford, 2nd ed. 10, 11

Wiener, Norbert. 1914. A contribution to the theory of relative position. *Proceedings of the Cambridge Philosophical Society* 17: 441–9. 10

Williams, Donald. 1951. The myth of passage. *Journal of Philosophy* 48: 457–72. 102, 125

Williamson, Timothy. 1998. Bare possibilia. *Erkenntnis* 48: 257–73. 82

——. 1999. Truthmakers and the converse Barcan formula. *Dialectica* 53: 253–70. 54

——. 2002. Necessary existents. In Anthony O'Hear, ed., *Logic, Thought and Language*, vol. 51 of *Royal Institute of Philosophy Supplement*, 233–51. Cambridge University Press, Cambridge. 82

——. 2006. Indicative versus subjunctive conditionals, congruential versus non-hyperintensional contexts. *Philosophical Issues* 16: 310–33. 42

Woods, Michael. 1976. Existence and tense. In Gareth Evans and John McDowell, eds. *Truth and Meaning*, 248–62. Clarendon Press, Oxford. 83

Zalta, Edward. 1988. *Intensional Logic and the Metaphysics of Intentionality*. The MIT Press, Cambridge, Mass. 65

Zimmerman, Dean. 1998. Temporary intrinsics and presentism. In Peter van Inwagen and Dean Zimmerman, eds., *Metaphysics: The Big Questions*, 206–19. Blackwell, Oxford. 94

——, ed. 2004. *Oxford Studies in Metaphysics*, vol. 1. Clarendon Press, Oxford. 90, 140, 145, 147, 148

——. 2008. The privileged present: Defending an 'A-theory' of time. In Theodore Sider, John Hawthorne, and Dean Zimmerman, eds., *Contemporary Debates in Metaphysics*. Blackwell, Cambridge, Mass. 87, 93

INDEX

Printed and bound by CPI Group (UK) Ltd, Croydon, CR0 4YY